ISBN 978-1-332-82415-1
PIBN 10066709

This book is a reproduction of an important historical work. Forgotten Books uses state-of-the-art technology to digitally reconstruct the work, preserving the original format whilst repairing imperfections present in the aged copy. In rare cases, an imperfection in the original, such as a blemish or missing page, may be replicated in our edition. We do, however, repair the vast majority of imperfections successfully; any imperfections that remain are intentionally left to preserve the state of such historical works.

English
Français
Deutsche
Italiano
Español
Português

www.forgottenbooks.com

Mythology Photography **Fiction**
Fishing Christianity **Art** Cooking
Essays Buddhism Freemasonry
Medicine **Biology** Music **Ancient**
Egypt Evolution Carpentry Physics
Dance Geology **Mathematics** Fitness
Shakespeare **Folklore** Yoga Marketing
Confidence Immortality Biographies
Poetry **Psychology** Witchcraft
Electronics Chemistry History **Law**
Accounting **Philosophy** Anthropology
Alchemy Drama Quantum Mechanics
Atheism Sexual Health **Ancient History**
Entrepreneurship Languages Sport
Paleontology Needlework Islam
Metaphysics Investment Archaeology
Parenting Statistics Criminology
Motivational

ON

RECENT IMPROVEMENTS

IN

WINDING MACHINERY,

BY JULES HAVREZ,

MINING ENGINEER.

TRANSLATED BY

R. F. MARTIN, C.E.

(Late Scholar of Trinity College, Cambridge.)

LONDON:

W. M. HUTCHINGS, 5, BOUVERIE-STREET, E.C.

1875.

TABLE OF CONTENTS.

PLATES.

INTRODUCTION.

STATEMENT OF THE OBJECT OF THIS PAPER.

THE tendency of pits is to become deeper, and of the quantities of coal drawn to become larger ; and, therefore, winding machinery is of continually increasing importance, and is for ever being perfected.

It is attracting particular attention in the Charleroi coal-field, where coal getting is now being carried on at a depth of 700 and even of 800 metres (765 to 875 yards.)

It is thus necessary, in order to have a large output, to run the cages in the shafts at speeds which approach those of locomotives, and to give the guides and the-headgear sufficient rigidity to stand the increased strains. All is minutely dependent upon the winding machinery ; and if we are compelled to adopt, instead of the old engines of Watt, double-cylinder engines, of 300 or 400 horse-power, and do not at the same time strengthen our pit guides and head-gear, we make a great error in design.

Moreover, in order to increase the output from our Neuville workings, without any stoppage, we found that we ought to put down an entirely fresh drawing pit, beside the present pit, whose old Watt engine, head-gear, and pit guides were correctly proportioned for each other, and might yet serve for sending down the men with new ropes, and for drawing the stone which had to be got out for the new workings.

We tried to adopt in our new pit the latest improvements which had been worked out in England, in Westphalia, and in Belgium, in winding machinery ; and it is the investigations that we undertook for this object which I am now going to detail.

We had the advantage of the scientific advice of M. J. Kraft, the chief engineer of John Cockerill's works at Seraing.

B

We took M. Kraft's opinion more particularly upon the best methods of applying steel ropes, either round or flat, and flat hemp ropes, to the depth and position of our new workings.

The problem which we had to work out involved the designing of a winding engine whose main shaft should not be farther from the axles of the pulley wheels than 15 metres (49·2 ft.), and which would produce, with the best possible results as regards consumption of fuel, and wear and tear of ropes, a large output of coal from a depth which must range from 460 to at least 700 metres (from 503 to 765 yards.)

It will be easily seen how important it is, in such a case, to reduce the dead weight of the cages and trams, by making them of steel; and to lighten the ropes (whose weight increases to a marvellous extent with the depth of the pit) by making them of the lightest possible material to give the proper strength, and by tapering them as much as is practicable.

It will appear presently that by employing cages and trams of steel, and round steel ropes, it is possible to obtain a *constant moment* of resistance with scroll drums of practical shape, even with a depth of 1000 metres (1093 yards).

It will appear, moreover, that by lightening the suspended weights, and tapering hemp ropes from hundred metres to hundred metres, or, better still, from twenty-five metres to twenty-five metres, we can make ropes sufficiently light to produce equilibrium with depths of 700 metres (765 yards), and with the radii of the drums, analogous to those which we now employ, and which produce equilibrium for depths of 500 metres (546 yards) only.

We can, in this case, employ winding engines driven with a constant grade of expansion ; or with an expansion which can only be varied at the end of the run and in changing the trams ; in a word, we may employ the most simple and the most economical form of winding engine.

This will show the intimate connection that exists between all the forms of winding engines, and the importance of considering them methodically.

We shall begin by examining the different forms of pit guides ; then of cages ; then of ropes, and the methods of balancing them ; lastly, of winding engines.

We shall then be able to see, in its true light, the improvements that have been recently produced in winding machinery, and the application of a variable cut-off, so as to proportion the power to the unequal resistance of the load.

PIT GUIDES.

A design for guides to draw a large quantity of coal must be, (1) strong, (2) easy to be maintained in good order, and to be replaced if the walling of the pit should come in, (3) arranged in two separate compartments, and so that one side can be worked at a time, in case the guides are torn out on the other side by an accident.

1. STRENGTH.—In order to combine these different requirements in a pit with arched walls, and with a longer axis of 3·60 metres (11·81 ft.) and a shorter axis of 2·40 metres (7·87 ft.) we divided it into two sides by an oak framing, shown in plate IX. figs. 1 and 2.*

Every metre in depth the pit is fitted with these "buntons." The two outside ones are 2·50 metres × 0·15 deep × 0·12 broad (98·42 in. × 5·89 in. × 4·71 in.) ; the middle one, which has more work to do, is 2·75 metres long × 0·20 deep × 0·15 broad (108·26 in. × 7·87 in. × 5·89 in.).

The rods themselves are 4 metres (13·12 ft.) long and 0·18 metres broad × 0·13 metres thick (7·08 in. × 5·11 in.). These are, it will be observed, unusually heavy. The large width was given to them so as to increase as much as possible the surface in contact with the buntons, and also to diminish the wear to which they are subjected by the friction of the cage-slides. It is well known that the slides wear the face and sides of the rods to an extent proportionate to the square of the speed of the cages. It is, therefore, advisable to give the rods as large a front surface as possible, to make them very good, and to prevent the wear upon the side faces from making the rods too thin. By means of buntons placed 1 metre (39·4 in.) apart, centre to centre, and rods as strong

---

* The modern English practice of invariably making shafts circular in section does not obtain in France and Belgium. They have oval pits, and pits made of four segments of circles, as in the present instance ; also square, rectangular, and polygonal shafts, are not uncommon. These latter three being often tubbed with wooden tubbing.—*Translator.*

as these, tied together by oak fishes of the same scantling as the rods, a system of guides is obtained of extraordinary strength ; and also the longer sides of the pit are strutted up and strengthened by the timbers which are fastened into them.

2. MAINTENANCE.—The outside buntons stand out from the shorter sides of the pit, so as not to weaken, unnecessarily, the masonry corners ; and also to allow of the rods being fixed by screws and bolts, so that they may be easily got at in spite of any settlement of the strata. There is, therefore, a distance of 0·35 metres (13·77 in.) between the walling of the pit and the back of the bunton.

The rods are not fastened directly on to the buntons, but have packings between each, of a thickness of 0·02 metres (0·78 in.), and a length and breadth equal to the depth of the bunton and breadth of the rod respectively. These packings give the rods, in a certain way, a thickness of 0·15 metre (5·9 in.) They can be taken out and replaced by others of a proper thickness, or even be taken out altogether in case of a settlement of the strata, so as to keep the rods accurately in the same vertical plane.

The outside rods are fastened to the buntons by bolts with projecting heads, screwed up at the back, so that they can always be easily tightened. The middle rods are fastened to the middle buntons by bolts with countersunk heads, in the manner shown below. The fishes of the rods are fixed with four bolts each. The screwed ends of the bolts are 0·02 metres (0·78 in.) in diameter.

We will now show the method adopted to render the rods of each side of the pit altogether independent of each other. It was devised by M. Charles Lambert, the engineer-in-chief, and it is thought that it accomplishes its object almost to perfection.

3. M. LAMBERT'S PLAN OF DIVIDING A DRAWING PIT INTO TWO COMPARTMENTS ALTOGETHER INDEPENDENT OF EACH OTHER.—The method is based upon the observed fact that, whenever accidents take place from cages coming out of the guides, it is the rods which are torn out of place, and the buntons are usually left intact.

M. Lambert thus, instead of having a double set of buntons in the middle of the pit, employs only one single set, and fastens to it two sets of rods. These are not put one at the back of the other, but are in two different vertical planes. Whenever, then, the rod belonging to one set is torn out by an accident the fellow rod of the other set remains untouched, because, though it is fastened to the same bunton, it has different bolts. It is to be remarked that each middle rod is fastened to the middle bunton better by this method, than if the rods were placed back to back, and thus required a bolt with a longer screwed end.

In fishing the rods with the wooden fishes, the only matter that need be attended to is that the joints of the two sets of rods do not come at the same place, but are, for instance, the one 1 metre beyond the other. If the fishes are cut of the same length as the distance between two buntons, they may be bolted to each rod with two bolts without sensibly decreasing the strength of the rods.

The plan devised by M. Lambert has the following advantages over the one in which a double set of buntons is fastened into the centre of the shaft.

1. The saving of one-half the large middle buntons.

2. The diminishing of the length of the longer axis of the shaft.

3. The strengthening of the whole timber, for the buntons are supported on both sides by the rods.

It is necessary, according to this plan to place the axes of the pulleys and the centres of the prop shafts in different planes on the two sides of the pit, and to make the props move rather farther forward on the one side than on the other.

It can, therefore, only be employed when the head gear is set up afresh. The system of guides, then, that we have described, combines economy with strength, and with ease in maintenance, and it divides the pit into two compartments, which are altogether independent of each other. All the pieces employed in its construction must be cut out beforehand, they can then be fixed in position without stoppage for 500 metres (546 yards) in depth. It is the best way to insure a perfectly vertical set of rods, and to render its erection economical as well as accurate.*

---

* This plan of putting in pit guides is very strong and solid, and presents several advantages. There are, however, some points which are contrary to our English practice in such matters; thus, the length of the rods, about 13 or 14 ft. is very small, and there is no solid diaphragm as is generally the

CAGES.

The most recent and the most sensible practice that has been introduced in the manufacture of cages, consists in the building them of steel, that is to say, of the strongest and lightest material known. It is in England, and particularly in Lancashire, that we saw this carried out, in 1866. On the Continent, in France and in Belgium, where iron cages are employed, a far higher proportion of dead weight to useful load is rendered necessary ; thus, in the work on collieries published recently by M. Burat, page 297, the following table is given :—

ANZIN.

	Weight.		Per centage.
	Kilogs.	Lbs.	
Two trams	420 ...	926·100 ...	35
Cage	1580 ...	3483·900 ...	53
Load of coal	900 ...	1984·500 ...	32
	2900 ...	6394·500 ...	100

BLANZY.

	Weight.		Per centage.
	Kilogs.	Lbs.	
Two trams	600 ...	1323·000 ...	19
Cage	1500 ...	3307·500 ...	49
Load of coal	1000 ...	2205·000 ...	32
	3100 ...	6835·500 ...	100

case in England, where guides are put in the middle of the pit. The following method will be found to produce a very good result. Four vertical 3-in. planks are spiked into the joints of the lining of the pit, so as to leave a 3-in. groove between them on each side of the pit. Into these two grooves the ends of 3-in planks, cut accurately to length, are slipped, and thus is formed a solid 3-in. diaphragm in the middle of the pit. On either side of this diaphragm one or more iron flat-footed rails are spiked, and their joints carefully fished with a fish underneath the bottom of the rail. These rail form the pit guides, and will last for many years without repairs.—*Translator.*

ANZIN.

	Weight.		Per centage.
	Kilogs.	Lbs.	
'Four trams	840	... 1852·200	... 19
Cage	1845	... 4068·225	... 41
Load of coal	1800	... 3969·000	... 40
	4485	... 9889·425	... 100

CHARLEROI.—NORTH.

	Weight.		Per centage.
	Kilogs.	Lbs.	
Four trams	580	... 1278·900	... 17
Cage	1480	... 3263·400	 43
Load of coal	1400	... 3087·000	... 40
	3460	... 7629·300	... 100

Thus, in *two-tram* cages, the useful load is about one-third of the total load : in *four-tram* cages the proportion is about two-fifths.

In Lancashire, on the other hand, the application of steel to the building of cages, and to the axles and ironwork of trams, has enabled them to lighten the proportion of dead weight to useful load in an astonishing manner ; as will be seen from the following table :—

CLIFTON HALL PIT, MANCHESTER.

	Weight.		Per centage.
	Kilogs.	Lbs.	
Four trams	628	... 1384·740	... 18
Cage (iron)	1270	... 2800·350	... 36
Useful load	1625	... 3583·125	... 46
Total	3503	... 7768·215	..., 100

California Pit, Wigan.

	Weight.		Per centage.
	Kilogs.	Lbs.	
Four trams	600	... 1323·000	... 22
Cage (steel, with two decks)	660	... 1455·300	... 24
Useful load	1520	... 3351·600	... 54
Total	2780	... 6129·900	... 100

Rose Bridge Pit, Wigan.

	Weight.		Per centage.
	Kilogs.	Lbs.	
Four trams	608	... 1340·640	... 20
Cage (steel, with two decks)	812	... 1790·460	... 27
Useful·load	1625	... 3583·125 ·	... 53
Total	3045	... 6714·225	... 100

East Canal Pit, Ince Hall.

	Weight.		Per centage.
	Kilogs.	Lbs.	
Six trams	600	... 1323·000	... 18
Cage (steel, with two decks)	1007	... 2220·435	... 30
Useful load	1830	... 4035·150	... 54
Total	3437	... 7578·585	... 100

We thus see that by employing steel in the building of cages the useful load exceeds the half of the total load; and that steel cages weigh less by one-half than iron ones.

It will appear, further on, how important it is to lessen the dead weight, so as to lighten the ropes, and increase the available force of the engines.

The Life of Steel Cages.—It must not be thought for a minute that because steel cages in Lancashire are light they are not stout and lasting.

Thus, the Ince Hall cages had been working when we visited the place, for four and a-half years, and had drawn

360 tons of coal and 50 tons of dirt, every day, from a depth of 540 metres (598 yards) at a speed of 7·70 metres (25·25 ft.) per second, and were even then in good case. At Rose Bridge, the steel cages, which weighed 812 kilogs. (1790 lbs.) had worked for four years, with a daily "turn" of 637 tons (coal and dirt together) ; and, in spite of a few breakages, their life might be estimated (so said the manager) at twice as long as this. These cages were, nevertheless, run at an average speed of 12 metres per second (39·36 ft.), and with their double decks had to stand the strains of "changing" very quickly indeed.

THE COST OF CAGES.—The steel cages at Ince Hall were built at the pit, and cost £32 each ; of this, £23 was for the rough steel itself.

The Rose Bridge cages cost £35 a-piece. We were so convinced of the advantage of using steel for cages, that we built for our own collieries some steel cages arranged for four trams, one over the other ; and although, *in iron,* they had weighed 1400 kilogs. (3087 lbs.) *in steel* they only weighed 1000 kilogs. (2205 lbs.) including the " Sibotte parachute."

The steel cages are lasting more than three years, while the iron cages were destroyed in less than half that time. We intend to replace these steel cages, which are at the present time not in a good state, by still lighter cages, weighing only from 850 to 900 kilogs. (1874 lbs. to 1984 lbs.) including the parachute. The weight of the " Sevrin" parachute is not more than 50 kilogs. (110 lbs.) Our cages have to stand great strains, on account of the double " change" which they have during the day, and we think, therefore, that it is well to increase their strength as much as possible by employing first-rate steel. Thus, the first cages were built of puddled steel supplied by the Montigny-sur-Sambre Company—and weighed 150 kilogs. (330 lbs.) more than the new cages, which will be built of Bessemer steel from Seraing. It is to be noted that four decks (one over the other) instead of two, necessarily make the cages heavier, and do not allow of their having the lightness of the English cages described above.

WEIGHT OF TRAMS.—We not only considered how the weight of the cages might be lessened, but tried to lighten the trams too. We, therefore, replaced their cast iron wheels with wrought-iron ones, made by the Couillet Company, and 0·30 metres (11·78 in.) in diameter, weighing 10 kilogs. (22 lbs.) each, and we also adopted axles and pedestals of Bessemer steel, made by M. Valere Mabille, of Morlanwelz. Each pair complete thus weighs only 29·6 kilogs. (65·25 lbs.), of which 9·5 kilogs. (20·94 lbs.) is for the axles and pedestals, and only cost 17s. 6d.*

It is to be remarked that the use of steel axles enables their weight to be diminished, and, therefore, also the diameters of the journals, which are as little as 0·03 metres (1·18 ins.) ; and this lessens the friction considerably. In this manner, our wooden trams, which hold about 4 hec. (14,126 cubic feet), weigh, empty, 190 kilogs. (418·87 lbs.) ; and their load of coal is 400 kilogs. (881 lbs.) ; so that the ratio of the dead weight to the useful load is not so much as one to two. M. Demanet, the engineer of the Esperance Colliery at Seraing, has even been able to reduce the dead weight of his trams more than this, by making their body of Bessemer steel. His tram, when well designed, only weighs 180 kilogs. (396 lbs.) for a load of 500 kilogs. (1102 lbs.) of coal ; that is a little more than one-third of the useful load ; their cost, however, is £7 5s. a-piece.

We intend to try some iron trams, which will hold 4 hec. (14,126 cubic feet), and weigh from 150 to 160 kilogs. (330 lbs. to 352 lbs.), so as to reduce the dead weight as much as possible.

The strains that the trams are subject to on the inclined planes have not enabled us to build wooden trams holding 4 hec. (14,126 cubic feet) as light as 150 kilogs. (330 lbs.), as they do in England. We have been obliged to increase the weight to 190 kilogs. (418 lbs.) in order to give the trams the necessary strength to stand the corresponding work to what they do in England ; but if we employ a material lighter and stronger than wood, we shall probably

---

* ·A great stride has been made in the last few months by the adoption of steel tram wheels. These are made of a mild "pot steel" and annealed carefully in an oven after they are cast. The result is one of the most tough and enduring of metals, for such a purpose, which has ever been produced ; and it is claimed for them that one third of the weight of the wheels may be saved, and that they last four times as long as cast iron ones. The cost of the metal at present is about three times that of cast iron.—*Translator.*

Pits.	Working depths.	Ropes.					
		Description.	Length.	Breadth.	Thickness or diameter.	Weight per metre.	Price per kilo.
California Pit.	Metres. 282	Round steel.	Metres. 343	Mm. —	Mm. 25	Kilo. 2·24	Fr. 1·50
	Yds. 308	—	Yds. 375	—	In. ·984	Lbs. 4·9	—
Clifton Hall Pit.	Metres. 388 and 486	Round steel.	Metres. 438	—	Mm. 28	Kilo. 2·48	1·50
	Yds. 424 and 531	—	Yds. 479	—	In. 1·1	Lbs. 5·5	—
Rose Bridge Pit.	Metres. 540	Flat steel.	Metres. 630	112·5 to 87·5	Mm. 19	Kilo. 5·15	1·875
	Yds. 590	—	Yds. 689	In. 4·4 to 3·4	In. ·748	Lbs. 11·4	—

be able to build them lighter without injuring their strength.

In the collieries of the Liege district, our weights are about as follow :—

The useful load as in England is rather more than half the total load.

	Weights in kilogs.	Lbs.	Per centage.
Four trams...	600 to 650	1323 to 1433	20 to 22
Cage	850 to 900	1874 to 1984	28 to 30
Useful load...	1600	3528	52
Total	3100	6725	100

DETAILS OF THE WEIGHTS DRAWN IN SOME ENGLISH PITS. —What we have just said about steel cages, shows the importance of what may be learnt from the practice of collieries in England, the land which is the pioneer of industrial progress. When we visited the English pits, in 1866, we were struck with the fact that the ropes were all made of metal, and, particularly, that they were round. Steel is used more and more for their ropes, particularly in the Lancashire coalfield. Contrary to our practice, the engines and the pulleys are so placed that the round ropes are in a capital position, both for their work and for lasting well. We have given in the following table the details of the winding gear and weights drawn in some of the most remarkable of the English pits (so far as this is concerned). We learn from M. Vilain, the manager of the Poisier collieries, and M. Depoitier, mining engineer, who visited, last year, the collieries in the North of England, that the use of round steel ropes and scroll drums was spreading more and more. They saw at a new sinking near Hetton a large drum to counterbalance round steel ropes, which were to draw 1000 tons of coal a day from a depth of 540 metres (589 yards.) The engines had two horizontal cylinders coupled—of 1·20 metres diameter and 1·80 stroke (48 in. in diameter, with 6 feet stroke). It was thought that by this means a thorough counterbalance and long life for the ropes would be combined. (*See table on folding page opposite.*)

ROPES.

ROUND STEEL ROPES.—The following table is compiled from the circulars of English rope makers.*

Name.	Diameter of rope. Mm.	Weight per metre. Kilogs.	Working load. Kilogs.	Breaking strain. Kilogs.	Ratio of working load to breaking strain.
Hartlepool Ropery Co.	25	2·24	4368	26,000	1 to 6
Wilkins and Co. ...	25	2·	4200	25,000	1 to 6
Newall	25·9	2·	3962	26,000	1 to 6·5
Do.	28	2·5	5486	36,000	1 to 6·5
Do.	30	3·	6096	40,000	1 to 6·5

Let us examine whether these loads are not exaggerated.

If the first rope were changed into a straight rod of steel, its weight might be decreased in the proportion of eight to nine, on account of its not being weakened by the torsion. It would then be equivalent to one weighing 1·99 kilogs. per metre (4·01 lbs. per yard), and this, divided by 7·8 kilogs. (the density) would give 255 square millimetres in section. Each square millimetre (0·001 in.) would, therefore, carry 17 kilogs.   Thus, in

Rope 1, the load per square millimetre is 17 kilogs. (37·485 lbs.)
    „ 2,   „   ·   18·4 „   (40·572 „)
    , 3,   17·4 „   (38·367 „)
    , 4,   19   „   (41·895 „)
    „ 5,   „   17·8 „   (39·249 „)

But according to the experiments made at Seraing (and these agree with the figures given by the principal makers), no one can safely trust steel to stand more than 14·3 kilogs. per square millimetre of section (31,530 lbs. to the square inch).   It will not then be safe to accept the English figures which would show that a rope weighing 1

* Instead of giving the English equivalents of the following French measures and weights, we have preferred to place the makers' cards in Appendix A, which see.

kilog. (2·205 lbs.) to the metre may have a working load of 2000 kilogs. (4410 lbs.)*

But let us revert to the data which we ourselves collected in England. The results are as follows :—

At Clifton Hall, a rope 1 kilog. (2·205 lbs.) to the metre (39·371 in.) would have a working load of 1700 kilogs. (3748·5 lbs.) This gives 14·9 kilogs. (32·854 lbs.) per square millimetre of section (32,854 lbs. per square inch.) At California Pit, a rope 1 kilog. per metre bears a mean load of 1500 kilogs., or 1 square millimetre of section bears a mean load of 13·1 kilogs. (28,885 lbs. per square inch.) This is not an exaggerated estimate, and we shall, therefore, adopt this as the basis of our calculation. It is to be noted that by taking 13 kilogs. per square millimetre, we are loading the ropes with one-eighth of the breaking weight, as given by the English makers ; and, moreover, if we calculate the ropes in this fashion for coal drawing, the working load will be one-seventh of their total strength when dirt is drawn instead of coal.

According to the experiments made by M. Demot in 1860, at Gosselies, before the mining commission on flat ropes, made of manganised steel, it was found that the square millimetre of steel wire had a breaking strain of from 95 to 107 kilogs. (209·4lbs. to 235·9lbs.) As the breakage took place without the slightest shock and in the most regular manner, whilst in drawing coals the ropes have to stand strains and oscillations and friction, the commission has taken only 80 kilogs. per square millimetre of section as breaking strain, and for working load, one-sixth of this strain, or say 13·3 kilogs. per square millimetre (29,326 lbs. per square inch).

According to M. Larivière, the manager of the Angers slate quarries, a steel wire of 2·1 millimetres (0·087 in.) diameter, weighing 0·021 kilogs. per metre has a breaking strain of 269 kilogs., so that the square millimetre (·001 square inch) will carry 77 kilogs. without breaking, and have a working load of 13 kilogs. (28·660 lbs.)

---

* We think that every one will allow that our knowledge of the properties of different classes of steel, and particularly of steel ropes, is imperfect; and that the different qualities of steel used by different makers, as well as our special advantages in England for obtaining good steel wire, will quite account for these differences.

ROUND STEEL ROPES OF UNIFORM SECTION.—In order to draw a total weight of 3100 kilogs. (6834·3 lbs.) we should have, according to these data, a parallel rope weighing 2700 kilogs. (5952·5 lbs.) or 3·86 kilogs. per running metre. We never saw them use, in England, ropes which were not parallel. This is to be explained by the shallow depths from which they draw, the average being not more than 300 metres (328 yards), and when the depths are much greater, as at Rosebridge, the English rope makers make their ropes with taper strands ; but it is not impossible to make ropes that taper from 100 metres to 100 metres by dropping out a steel wire every now and then.

ROUND STEEL ROPES TAPERING FROM 100 METRES TO 100 METRES.—Let us consider a round steel rope tapering every 100 metres and 700 metres long (766 yards), and whose total working load shall be 3100 kilogs. (6834·3 lbs.), of which 1600 kilogs. (3527·4 lbs.) are useful load ; and let us assume that a rope 1 kilog. per metre will bear 1500 kilogs. (3206·9 lbs.) at the top of each length of 100 metres (328 ft. 1 in.).

We shall then have :—

Length. Metres.	Weight per metre run.	Weight of each length.	Difference of weights of lengths.	2nd diff.	3rd diff.
100 ...	2·218 ...	221·8			
			15·7		
100 ...	2·373 ...	237·3		1·3	
			17·0		0·1
100 ...	2·543 ...	254·3		1·2	
			18·2		0·0
100 ...	2·725 ...	272·5		1·2	
			19·4		0·0
100 ...	2·919 ...	291·9		1·2	
			20·6		0·0
100 ...	3·125 ...	312·5		1·2	
			21·8		
100 ...	3·343 ...	334·3			

700 m. at 2·750   = 1924·6

In order to obtain a general expression for the weight of round steel ropes which taper from 100 metres to 100 metres, and of a length of m hundred metres, it has been pointed out by Mr. P. Havrez that the 4th difference in the variations of the weights of the successive sections being zero, the usual formula of interpolation may be applied. This, arranged in order of ascending powers of m, is :—

$$\phi\,(\kappa + m\Delta\kappa) = \phi\kappa + \left(\frac{\Delta\phi\kappa}{1} - \frac{\Delta^2\phi\kappa}{1\cdot2} + \frac{2\Delta^3\phi\kappa}{1\cdot2\cdot3}\right)m$$

$$+ \left(\frac{\Delta^2\phi\kappa}{1\cdot2} - \frac{3\Delta^3\phi\kappa}{1\cdot2\cdot3}\right)m^2 + \frac{\Delta^3\phi\kappa}{1\cdot2\cdot3}\,m^3.$$

Here $\phi\kappa$ = Weight of rope no metres long = 0.

$\Delta\,\phi\kappa$ = 222 kilogs.　(Approximate.)

$\Delta^2\phi\kappa$ = 15·7 kilogs.

$\Delta^3\phi\kappa$ = 1·3 kilogs.

$\Delta^4\phi\kappa$ = 0.

∴ Weight of m hundred metres of steel rope

$$= \left(\frac{222}{1} - \frac{15\cdot7}{2} + \frac{1\cdot3}{3}\right)m + \left(\frac{15\cdot7}{2} - \frac{1\cdot3}{2}\right)m^2 + \frac{1\cdot3}{6}\,m^3$$

$$= 214\,m + 7\cdot2\,m^2 + 0\cdot2\,m^3.$$

If $m = 1$　$P_1 = 214 + 7\cdot2 + 0\cdot2 = 221\cdot4$ kilogs.

$m = 2$　$P_2 = 428 + 28\cdot8 + 1\cdot6 = 458\cdot4$ kilogs.

$m = 7$　$P_7 = 1498 + 352\cdot8 + 68\cdot6 = 1920$ kilogs.

These calculations agree closely with the figures given above.

Thus, the increase of the weight of ropes begins to be very great, on account of its involving the 3rd power of the depth, when the depth is more than 700 metres, and their thickness must increase in the same manner.

It will thus be seen that to lighten ropes as much as possible, and thus draw coals from great depths, it is a matter of necessity to taper the ropes : and that the sections increase rapidly in weight as their differences increase progressively. For as the depth is increased the size of the last sections of the rope become enormous, and it is, therefore, of great importance to make the load at the end of the rope as small as possible, as this is the first term in this series.

ROPES TAPERING CONTINUOUSLY FROM END TO END.—It may be interesting to work out the weight of a rope which tapers continuously. Now in iron and steel it will be theoretically necessary, for this to make the rope with each wire tapering, and the wire-drawing must be considered and varied in order to do this. For this purpose, we must make great strides in the manufacture of wire, but let us see what will be the results of that in the lightening of ropes.

1. What is the necessary section at any point of a rope in order that it may always be strong enough to carry its own weight.

Let a be the section of the rope at its lower end.

Q = the load.

r = the strength of a unit of section of the rope.

y = the specific gravity of the material of the rope.

$\delta\kappa$ = a small element of length of the rope.

S = the section of the rope at the distance κ from the lower end.

Then we have, Strength of rope = weight at every point, or,

$$S\,r = Q + \text{weight of rope below that point}$$

$$= Q + \int_0^x S y \delta\kappa$$

(For $S y \delta\kappa$ is the weight of an element of length),

$$\therefore \quad r\,\delta S = S y \delta\kappa$$

$$\text{or} \quad \frac{\delta S}{S} = \frac{y}{r}\,\delta\kappa$$

$$\therefore \text{ integrating log. } S + c = \frac{y}{r}\,\kappa \quad \ldots \ldots \ldots \quad (1)$$

But when

$$\kappa = 0 \quad S = a \text{ and } Q = ar \text{ or } a = \frac{Q}{r}$$

and at the lower end (1) becomes

$$\log. a + c = 0; \quad \text{or } c = -\log. a = -\log. \frac{Q}{r};$$

$$\therefore \text{ Substituting log. } S - \log. \frac{Q}{r} = \frac{y\kappa}{r}$$

$$\text{or log. } \frac{Sr}{Q} = \frac{y\kappa}{r},$$

whence

$$\frac{Sr}{Q} = e^{\frac{y\kappa}{r}} \text{ and } S = \frac{Q}{r}\,e^{\frac{y\kappa}{r}} \quad \ldots \ldots \ldots \quad (2)$$

(2.) What is the volume of a continually tapering rope ?

The volume V will be $\int_0^l S\delta\kappa$.

Where l is the total length of the rope, and, substituting the value of S from (2)

$$V = \int_0^l \frac{Q}{r}\,e^{\frac{y\kappa}{r}}\,\delta\kappa = \frac{Q}{r}\int_0^l e^{\frac{y\kappa}{r}}\,\delta\kappa$$

To integrate this, let us write it in the form :—

$$V = \frac{Qr}{ry}\int_0^l \frac{y}{r}\,e^{\frac{y\kappa}{r}}\,\delta\kappa$$

$$= \frac{Q}{r} \frac{r}{y} \int_0^l e^{\frac{yx}{r}} \delta\left(\frac{yx}{r}\right)$$

Now, it will not be forgotten that the integral of

$$e^z = \frac{e^z}{\log. e} = e^z$$

So the integral of the above expression will be

$$V = \frac{Q}{y} e^{\frac{yx}{r}} \text{ between limits } l \text{ and } o$$

$$= \frac{Q}{y}\left(e^{\frac{yl}{r}} - e^0 \right) = \frac{Q}{y}\left(e^{\frac{yl}{r}} - 1 \right) \quad \dots \dots (3)$$

(3.) What is the actual weight of such a rope ?

This will be, of course, V (the volume) multiplied by the density y ; and

$$Vy = Q\left(e^{\frac{yl}{r}} - 1 \right)$$

In the case of round steel ropes, above considered,

Q = 3100 kilogs.
y = 7·8 kilogs. (the density).
l = length of rope.
r = 131,580 (the strength).
e = 2·718.

$\therefore e^{\frac{yl}{r}}$ will $= 2·718^{\frac{7\cdot8 \times 7000}{131580}} = 2·718^{0\cdot415}$.

Taking logs. log. $e^{\frac{yl}{r}} = 0·415 \times$ log. $2·718 = 0·415 \times 0·4343.$
$= 0·18023$

and $e^{\frac{yl}{r}} = 1·5144$

Thus P $= $ Q $(1·5144 - 1) = 0·5144$ Q
$= 0·5144 \times 3100$
$= 1594$ kilogrammes.

And a round steel rope of logarithmic form (as shown by the equation) would weigh only 1594 kilogs. ; or 371 less than a rope which tapers only from 100 metres to 100 metres, and 1106 kilogs. less than a rope of uniform section.

FLAT STEEL ROPES.

The following data are extracted from the cards of English rope makers :—*

Manufacturer's name.	Sizes of ropes. Centimetres.		Weight per running metre.	Working loads.	Reduced working load for a rope weighing 1 kilog. to the metre.	Ratio of working load to breaking strain.
	Breadth.	Thick-ness.	Kilogs.	Kilogs.		
Newall ...	5·0	1·25	2·5	3200	1280	1 to 8·95
Do. ...	6·8	0·94	3·71	5000	1340	1 to 9
Do. ...	8·7	0·94	5·0	6800	1360	1 to 8·88
Hartlepool Ropery	5·3	1·25	2·5	3000	1200	1 to 9
Do.	6·8	1·56	4·0	5000	1250	1 to 9
Do.	8·7	1·56	5·0	5500	1100	1 to 8
Wilkins & Weatherly	5·0	1·6	2·5	3200	1280	1 to 8·1
Do.	7·5	1·6	4·0	5400	1350	1 to 7·5
Do.	9·1	1·6	4·8	6000	1340	1 to 7·7
John Shaw	6·1	1·2	2·98	3454	1150	1 to 9
Do.	8·3	1·6	3·96	5130	1300	1 to 8·1
Do.	9·0	1·6	4·46	5993	1340	1 to 7·8

It will be seen that, according to this, the working load is only from one-eighth to one-ninth of the breaking strain, whilst for round ropes it amounts to one-sixth. This is because the different ropes (of which the flat rope is made) are never perfectly united in spite of the stitching which fastens them together.

Thus, while a round steel rope, weighing 1 kilog. per metre, would carry, according to the average of English makers, 2000 kilogs. working load, a flat steel rope, 1 kilog. per metre, would only bear the following loads :—

* Instead of giving the English equivalents of the following French measures and weights, we have preferred to place the makers' cards in Appendix A, which see.

According to Newall 1300
,, Hartlepool Co. 1200 } Average, 1270 kilogs.
,, Wilkins & Co. 1320 } (2800 lbs.)
,, Shaw 1260

If, then (as the English makers assert) a flat steel rope weighing 1 kilog. per metre will carry well, 1200 kilogs., it appears that the ratio of the working loads of a round and flat steel rope is 8 to 10.*

And as we found that a round steel rope would bear a working load of 13·158 kilogs. per square millimetre, so we should have for a flat steel rope 10·5264 kilogs. per square millimetre (23,210 lbs. per square inch.) Now let us see what it is in practice. At Rose Bridge Pit, a flat tapering steel rope 540 metres in length, and weighing, on the average, 5·15 kilogs. per running metre would have on its upper end the following load :—

Coal............................ 1625
Cage 812
Trams 608
Rope 2781
 ———
Total .... 5826 kilogs.

To carry this load we should have to use a rigid steel bar of 582·6 square millimetres section, since each square millimetre will support 10 kilogs. This bar would weigh 582·6 × 7·8 or 4·544 kilogs. But, on account of the weakening due to the twist of the rope, this must be increased in the ratio of 8 to 9, or ought to be 5·11 kilogs.

It will thus be seen that at Rose Bridge they do not even strain the rope up to 10 kilogs. per square millimetre (22,050 lbs. per square inch). We may, therefore, assume that a flat steel rope ought not to carry more than 10 kilogs. per square millimetre of section (we adopt this new variation of the unit on account of the great simplification of the calculations). And let us now calculate what will be the working load of a rope weighing 1 kilog. per running metre (2 lbs. per yard) according to this result. The section S of a rope weighing 1 kilog. to the metre will be determined by the equation—

* This would be 1200 to 2000 or 6 to 10, not 8 to 10, or 7·894 kilogs. per square millimetre.—*Translator.*

S (square millimetres) $= \frac{8}{9} \times \frac{1000}{7\cdot8} = 113\cdot9$ (square milli-metres) (the ratio $\frac{8}{9}$ representing the increase due to the torsion). Thus, a rope weighing 1 kilog. to the metre has a section of 113·9 square millimetres, and as each square milli-metre can carry a working load of 10 kilogs., the total working load will be 1139 kilogs. or, adding one in the units' place to produce even figures, 1140 kilogs. (2513 lbs.)

We shall, therefore, assume in our calculations that a flat steel rope weighing 1 kilog. to the metre will bear a working load of 1140 kilogs. (*i. e.* a rope weighing 2 lbs. to the yard will bear a working load of 2513 lbs.)

FLAT STEEL ROPE OF UNIFORM SECTION.—It will follow at once that, in order to draw a total load of 9100 kilogs. from a depth of 700 metres (20,065 lbs. from a depth of 766 yards), we should have to employ a parallel rope of 4932 kilogs. weight (10,875 lbs.) or 7·045 kilogs. per running metre (14·198 lbs. to the yard). For the total load which the topmost metre of the rope has to stand is 9032 kilogs. (19,915 lbs.); and dividing this load by the load (1140 kilogs.) which a rope 1 kilog. to the metre can carry, we find that we must give to this metre a weight of 7·045 kilogs. (15·53 lbs.) per running metre.

FLAT STEEL ROPE TAPERING FROM 100 METRES TO 100 METRES.—The following table gives the details of a flat steel rope 700 metres long, tapering each 100 metres, and suffi-ciently strong to carry a working load of 3100 kilogs. (6835 lbs.) at its lower end :—

Lengths.	Weights per running metre.		Weights of each length.	Differences.
Metres.	Kilogs.	Lbs.	Kilogs.	
100 ...	2·982 ...	6·575 ...	298·2	
100 ...	3·270 ...	7·210 ...	327·0	28·8
100 ...	3·584 ...	7·902 ...	358·4	31·4
100 ...	3·928 ...	8·661 ...	392·8	34·4
100 ...	4·304 ...	9·490 ...	430·4	37·6
100 ...	4·715 ...	10·396 ...	471·5	41·1
100 ...	5·165 ...	11·388 ...	516·5	45·0
700 ...	3·993 ...	8·804 ...	2794·8	

The formula of interpolation (before made use of) will immediately give an expression for the weight of a rope, as above, whose length is one hundred metres.

$$\text{For } \phi \left(\kappa + m \, \Delta \, \kappa \right) = \phi \kappa + \frac{m}{1} \Delta \, \phi \kappa \cdot + \frac{m \cdot \overline{m-1}}{1 \cdot 2} \Delta^2 \phi \kappa + \; .$$

$$\frac{m \cdot \overline{m-1} \; \overline{m-2}}{1 \cdot 2 \cdot 3} \Delta^3 \phi \kappa.$$

Or the weight

$$(0 + m. \text{ hundreds}) = 0 + \frac{m}{1} \, 298 + \frac{m \, \overline{m-1}}{1 \cdot 2} \, 29 + \frac{m \, \overline{m-1} \; \overline{m-2}}{1 \cdot 2 \cdot 3} \, 3$$

For the first difference between the weights of the sections of the rope is 298, the difference of the differences, 29, and the difference, again, of these, 3 ;

∴ Weight of m hundred metres or P

$$= \left(298 - \frac{29}{2} + \frac{3}{3} \right) m + \left(\frac{29}{2} - \frac{3}{2} \right) m^2 + \frac{m^3}{2}$$

$$P = 285 \, m + 13 \, m^2 + 0 \cdot 5 \, m^3$$

If $m = 1$ P $= 298 \cdot 5$ kilogs. (658 lbs.) .
  $m = 2$ P $= 626$,,   (1380 ,,)
  $m = 3$ P $= 785 \cdot 5$,,   (1732 ,,)

  $m = 6$ P $= 2285$,,   (5038 lbs.)
  $m = 7$ P $= 2803$,,   (6181 ,,) which figures are practically identical with those given above.

Now, comparing the formula for a round with a flat steel rope, we must observe that in the former case the co-efficient of m^2 is one-half, and that of m^3 $\frac{1}{2 \cdot 5}$ of the corresponding co-efficient in the latter case.

The greater the value of m, the depth, the more important is it to adopt a round rope instead of a flat one.

Thus, the weight for small depths varies according to m and m^2, but above 600 metres, far more, on account of the term involving m^3. And for great depths of 700 and 1000 metres (765·5 and 1093·6 yards) the weight of the rope increases rapidly with the cube of the depth. Thus, for 600 metres this term adds 108 kilogs. to the weight (238 lbs.) and for 700 metres it adds 172 kilogs. (379 lbs.) This law is also true in the case of a uniformly tapering rope, only it will make the rope rather too strong, and this is a good fault.

FLAT STEEL ROPE CONTINUOUSLY TAPERING.—What is the weight of a flat steel rope with a continuous taper ?

It is given by the formula :—

$$P = Q \left(e^{\frac{yl}{r}} - 1 \right)$$

in which $y = 7\cdot8$ $e = 2\cdot71828$
           $r = 100,000.$ $Q = 3100$
           $l = 7000.$

$$\therefore \quad e^{\frac{yl}{r}} = 2\cdot71828^{0\cdot546}$$

and log. $e^{\frac{yr}{r}} = 0\cdot546$ log. $2\cdot71828$
                 $= 0\cdot546 \times 0\cdot4343$
                 $= 0\cdot237128$

$$\therefore \quad e^{\frac{yl}{r}} = 1\cdot726$$

Whence P $= 3100 \times \cdot726 = 2251$ kilogs. (4963 lbs.)

As, therefore, the saving of weight is considerable, it will be important to try and manufacture steel ropes with a continuous taper.

IRON WIRE ROPES.

Iron wire ropes are more and more, every year, in England, being replaced by ropes of mild steel, seeing that a steel rope is to be considered only as the perfect form of metallic rope. The faults of metallic ropes (slight flexibility, difficulty in detecting bad places, too slight thickness in the rope to make a good counterbalance by the increase in the diameters as the rope rolls up) are nearly the same whether steel or iron is employed. We shall, therefore, not place any stress upon the details of *iron* ropes, seeing that steel may be made to replace it with advantage. Moreover, the comparisons of ropes made of metals, and of textile substances (such as aloes or hemp), which we are now going to begin, should, to be conclusive, be based upon the most perfect form of metal rope, that is to say, a steel rope, and upon the aloes rope which we use on the Continent.*

* It is a remarkable fact that the fibre which is apparently now used universally for hempen pit ropes in Belgium and the north of France should be so little known amongst miners in England. The history of the term " aloes " as applied to these ropes is curious. The *real aloe* is a plant of doubtful origin, seldom met with in this part of the world, though now widely spread in warm countries, especially in Africa, and known chiefly from the bitter medicine which is made from it. The so-called *Mexican aloe* (*Agave Americana*), which is in reality not an *aloe* at all, but an *amaryllid*, is now commonly known by that name on account of its resemblance to the plant

ROPES OF ALOES.

It is well known that aloes form the best fibre for the manufacture of pit-ropes, and that the use of hemp has now been given up for some time.

The following data have been furnished to us by the firm of Vertongen and Goens, of Termonde.

The breaking strain of aloes is 6 kilogs. per square millimetre of section (13,230 lbs. per square inch.) The working load is 0·75 kilogs. per square millimetre (1654 lbs. per square inch), so that the ropes are worked to one-eighth of their breaking strain.

It is to be noted that if they are worked up to one-eighth of the breaking strain in drawing coal, it will be increased up to about one-seventh in drawing stone and dirt. The weight of a cubic decimetre of aloes' fibre, supposing it to be solid, is about 1 kilog. (62·43 lbs. to the cubic foot.) So each metre of aloes rope which weighs 1 kilog., and, therefore, has 1000 square millimetres of section, or 10 square centimetres, would carry a working load of 750 kilogs. (A rope weighing 2·015 lbs. to the yard will have a section of one square inch, and carry a working load of 1653 lbs.) We will now calculate the weights of aloes ropes (1) if the section be uniform, (2) if the rope taper from 100 metres to 100 metres, (3) if the rope taper continuously, and is in each case fit to draw a total load of 3100 kilogs. from a depth of 700 metres (6835 lbs. from a depth of 766 yards.)

ALOES ROPE OF UNIFORM SECTION.—This rope must weigh 43,400 kilogs. or 62 kilogs. per running metre (95,697 lbs. or 120 lbs. per yard.) For the total load which the topmost

mentioned above. Its long fleshy leaves yield, when properly macerated, an exceedingly tough fibre, which is used in making nets and small articles of a similar nature. Hence, the term *aloes* has become so well known by French rope-makers. The word, however, is now used, in a much more general sense, to denote *various sorts of fibres used for cordage other than hemp.* The best of these *manila* is the fibre of the *Musa textilis*, a species of *plantain*, and not, in any way, an *aloe.* This is the fibre mentioned in the text. Among the other materials embraced in the general name *aloes* are, the Spanish esparto grass, used for ships' warps, and jute, which is chiefly used for sacking and inferior cord. It is to be noted that flat pit ropes made of *manila coated with hemp,* so that the outside covering may retain the tar, have been successfully used in England for some years.—*Translator.*

mètre of the rope has to carry is 46,500 kilogs. (102,530 lbs.) ; and if this be divided by 750 kilogs., which a rope weighing 1 kilog. per mètre will carry, we find that the topmost metre must weigh 62 kilogs. We must observe here that an aloes rope of uniform section would break from its own weight if reaching to the bottom of a pit 750 metres (820 yards) in depth.

ALOES ROPE TAPERED FROM 100 METRES TO 100 METRES. —An aloes rope, according to the conditions just detailed, would be made according to the following table :—

Lengths. Metres.	Weight per metre. Kilogs.	Weight of sections. Kilogs.	Differences between weights.
100	4·77	477·	
100	5.503	550·3	73·3
100	6·35	635·	84·7
100	7·327	732·7	97·7
100	8·454	845·4	112·7
100	9·754	975·4	130·0
100	11·254	1125·4	150·0
700	7·63	5341·7	

In a general form the weight of m hundred metres of an aloes rope $= 443\cdot7\,m + 31\cdot8\,m + 1\cdot4\,m^3 + 0\cdot08\,m^4$ for applying as before the method of differences :—

$$\text{For we find } \Delta\,\varphi\kappa = 477\cdot$$
$$\Delta^2\varphi\kappa = 73\cdot3$$
$$\Delta^3\varphi\kappa = 11\cdot4$$
$$\Delta^4\varphi\kappa = 2\cdot \text{ (constant.)}$$
$$* \quad \Delta^5\varphi\kappa = 0$$

* It will, I think, be found, if these figures be gone over carefully, that the complete statement of the formula is—

$$\Delta\,\varphi\kappa = 477\cdot$$
$$\Delta^2\varphi\kappa = 73\cdot3$$
$$\Delta^3\varphi\kappa = 11\cdot4$$
$$\Delta^4\varphi\kappa = 1\cdot6$$
$$\Delta^5\varphi\kappa = \cdot4 \text{ (constant.)}$$
$$\Delta^6\varphi\kappa = \cdot0 \quad \text{,,}$$

The effect will be to add one more term to the co-efficients of m, m^2, m^3, and m^4, and to introduce a term at the end $\cdot003\,m^5$. The result, however, for a depth of 700 metres (when $m = 7$) is strangely near that given in the text, the weight being 5331·7 kilogs. It is, therefore, worth while only to mention it in passing.—*Translator.*

And the formula before employed will give—

$$\varphi\,(\kappa + m\,\Delta\,\kappa) = \varphi\,\kappa + \left(\frac{\Delta\,\varphi\,\kappa}{1} - \frac{\Delta^2\,\varphi\,\kappa}{1.2} + \frac{2\,\Delta^3\,\varphi\,\kappa}{1.2.3} - \frac{6\,\Delta^4\,\varphi\,\kappa}{1.2.3.4}\right) m$$
$$+ \left(\frac{\Delta^2\,\varphi\,\kappa}{1.2} - 3\,\frac{\Delta^3\,\varphi\,\kappa}{1.2.3} + 11\,\frac{\Delta^4\,\varphi\,\kappa}{1.2.3.4}\right) m^2$$
$$+ \left(\frac{\Delta^3\,\varphi\,\kappa}{1.2.3} - 6\,\frac{\Delta^4\,\varphi\,\kappa}{1.2.3.4}\right) m^3$$
$$+ \left(\frac{\Delta^4\,\varphi\,\kappa}{1.2.3.4}\right) m^4$$

omitting all subsequent terms because $\Delta^5\,\varphi\,\kappa = 0$.

Substituting the values of $\Delta\,\varphi\,\kappa$, &c., the weight of $(0 + m$ hundred metres of rope)

$$= 0 + \left(477 - \frac{73\cdot3}{1.2} + \frac{2 \times 11\cdot4}{1.2.3} - \frac{6 \times 2}{1\cdot2\cdot3\cdot4}\right) m$$
$$+ \left(\frac{73\cdot3}{1.2} - \frac{3 \times 11\cdot4}{1.2.3} + \frac{11 \times 2}{1.2.3.4}\right) m^2$$
$$+ \left(\frac{11\cdot4}{1.2.3} - \frac{6 \times 2}{1.2.3.4}\right) m^3$$
$$+ \frac{2}{1.2.3.4}\ m^4$$

∴ The weight of m hundred metres of rope
$$= 443\cdot7\,m + 31\cdot8\,m^2 + 1\cdot4\,m^3 + 0\cdot08\,m^4$$
If $m = 1$ $P = 443\cdot7 + 31\cdot8 + 1\cdot4 + 0\cdot08 = 477$
$m = 2$ $P = 887\cdot4 + 127\cdot2 + 11\cdot2 + 1\cdot3 = 1027\cdot1$
. . . .
$m = 7$ $P = 3105\cdot9 + 1558\cdot2 + 480\cdot2 + 192\cdot8 = 5337\cdot1.$

It is evident from inspection that the co-efficients which have the greatest effect upon the weight are those of m^3 and afterwards of m^4 as the depth increases. Thus, it is only about at 700 metres that the co-efficient of m^4 adds 200 kilogs. to the weight of the rope.

We may, then, assume that the weight of an aloes rope 700 metres in length, as above, will be 5340 kilogs. (11,748 lbs.)

THEORETICAL FORM OF TAPERING ALOES ROPE.—Let us substitute in the formula

$$P = Q\,\left(e^{\frac{y\,l}{r}} - 1\right) \quad Q = 3100 \text{ kilogs.}$$
$$l = 7000 \quad\text{,,}$$
$$e = 2\cdot71828\text{,,}$$
$$y = 1 \quad\text{,,}$$
$$r = 7500 \quad\text{,, per decimetre.}$$

then $P = 3100\left(2\cdot71828^{\frac{1 \times 7000}{7500}} - 1\right)$

Taking logs. as log. $e = \cdot4343$

log. $e^{\cdot9333} = 0\cdot9333 \times \cdot4343 = 0\cdot40533$

∴ $e^{\cdot9333} = 2\cdot543$

and P = 3100 (2·543 − 1) = 4783 kilos.

or, on an average, 7 kilos. per metre (14·1 lbs. per yard.)

A specimen of this sort of rope was exhibited at the Paris exhibition which was manufactured by Messrs. Stievenard, Cambier and Son. It was tapered in a length of 25 metres from ·30 metre at one end down to ·18 at the other. I have received the following communication from the makers on this subject.

. "The manufacture of ropes on this plan is very difficult. In order to be really well done, a special form of apparatus is required, which only two or three makers on the Continent possess at present. We have here a machine which works quite alone, without needing any attention, except the replacing of the reels of yarn, and the taking away of the ropes as fast as they are-spun.

" The machine is 9 metres in height (29·5 ft.), and 5 metres in circumference. It will.turn out 1000 metres in a day, and takes up a space of only 10 metres square altogether, the machine taking up half of this space, and could make 10,000 or 100,000 metres of uniform section or tapering, all in one length, and without any splice, which is a most important matter. From the first metre to the thousandth the torsion is always the same per metre. The yarns, which are drawn down in a vertical direction, and will only draw out at a certain tension, are, in consequence, equally stretched, and all, without exception, carry the same load.

" According to the apparatus usually employed, each tapering strand is composed of three sets of yarns. Each of these sets which continually tapers is twisted along a rope-walk, and the smallest portion perforce receives a stronger twist than the rest. This is the first disadvantage. Moreover, it follows that the sets of yarns being placed horizontally, some are more stretched than others, and thus there is a loss of strength. Then, when they come to lay up the rope, that is to say, to lay together the three strands, to make the round rope, this operation again produces the same objection of a sharper twist in the weaker section. Most rope-makers, in order to get over some of these important objections, are obliged to make of each tapering section a separate rope, and to join up each section to the next as well as they can by a splice. If there are four sections there are three or four splices, which is a source of danger, is only like an old rope, produces an unequal tension in the different strands, an abrupt transition from one section to another, and tends, moreover, to untwist the rope. In any case, the result is bad, and is shown by frequent repairs, by a notably shorter life of rope, and sometimes actually by fracture. This will explain to you why in the Charleroi and Liége coalfields we find that the manufacture of ropes with tapering sections is very difficult and expensive, and gives bad and dangerous results.

"Our machine, which is of a considerable size, not only finishes the rope without any splice, but also makes cable-laid ropes gradually tapering so as to pass from one section to another by insensible degrees. The tension, which ought to vary as the size, in each portion of the rope, is regulated at pleasure by the gearing. You will have been able to appreciate by our specimen rope in the exhibition the remarkable results which we have obtained and which any rope-maker would declare impossible; in the short length of 25 metres (82 ft.) we have made it taper from a circumference of ·30 metre (11·8 in.) to ·18 metre (7 in.), without a single splice, without any untwisting of the eight strands, and by a decrease of strength so gradual as to be by inappreciable degrees. Much more, therefore, can we obtain results such as these in longer ropes.

"Below are the different sections of rope which we manufacture.

				Circumference.	
				Metre.	In.
1st length of 100 metres	......	·29 =	11·41
2nd ,, 100 ,,	......	·28	11·02
3rd ,, 50 ,,	......	·27	10·63
4th ,, 50 ,,	......	·26	10·23
5th ,, 50 ,,	......	·25	9·84
6th ,, 50 ,,	......	·24	9·44
7th ,, 50 ,,	......	·23	9·05
8th ,, 50 ,,	......	·22	8·66
9th ,, 50 ,,	......	·21	8·26
10th ,, 50 ,,	......	·19	7·48

Total...... 600 metres."

Messrs. Vertongen and Goëns, of Harmignies, manufacture, also, aloes ropes tapering in lengths of 50 metres, and even in lengths of 25 metres ; they can even push the amount of taper to its theoretical limits.

We give on page 29, in a tabular form, the weights and prices of different kinds of ropes 700 metres (766 yards) long.

It follows, from this table,

(1.) That it is of the highest importance to give our ropes a continuous taper. As for aloes, we may consider the question settled. In the present condition of the manufacture of steel ropes, one can hardly hope to get them tapering more frequently than every hundred metres. In order, however, to push this tapering still further, it will be necessary to make them of smaller wires, and to leave one out as often as possible.

(2.) That for a depth of 700 metres an aloes rope of a continuous taper is again about two and a-half times heavier, and more costly, than a round steel rope which tapers each 100 metres, and one and a-half times heavier and more costly than a flat steel rope of a similar make.

(3.) That when we come to be able to make constantly tapering metallic ropes, round steel ropes 700 metres long are nearly three times as light and more economical than continuously tapering aloes ropes, and flat steel ropes are twice as light and economical as aloes ropes.

Let us form a similar table for 1000 metres (1094 yards) in depth.

The weight of tapering ropes will be given by the general tables on page 30, for a load of 3100 kilogs. (6835 lbs.) and to lift from a depth of *m hundred metres*.

It is evident from what goes before—

(1.) That ropes of uniform section, even round steel ropes, have to be put out of the question when the depth is as great as 1000 metres (1094 yards).

(2.) That a flat steel rope tapered every 100 metres will be one and a-half times heavier than a round steel rope. The difference of the weights, if the rope be 1600 metres (1750 yards) long, will be exactly the whole useful load. The continuously tapering aloes rope is about three times as heavy as a round steel rope tapered every hundred metres. The fact that round steel ropes proved to be so much superior to

	Total weight of rope.			Least weight of rope per metre.			Price of rope.		
	Round steel rope.	Flat steel rope.	Aloes rope.	Round steel rope.	Flat steel rope.	Aloes rope.	Round steel rope.	Flat steel rope.	Aloes rope.
Rope of uniform section.	Kilogs. 2,700 / Lbs. 5,953	Kilogs. 4,932 / Lbs. 10,875	Kilogs. 43,400 / Lbs. 95,697	Kilogs. 3·86 / Lbs. 8·51	Kilogs. 7·04 / Lbs. 15·5	Kilogs. 62·00 / Lbs. 136·71	Fr. 4,050 / £ 162	Fr. 8,877 / £ 355	Fr. 69,440 / £ 2778
Rope tapering each 100 metres.	Kilogs. 1,925 / Lbs. 4,244	Kilogs. 2,797 / Lbs. 6,167	Kilogs. 5,341 / Lbs. 11,776	Kilogs. 2·75 / Lbs. 6·06	Kilogs. 3·99 / Lbs. 8·79	Kilogs. 7·63 / Lbs. 16·8	Fr. 2,887 / £ 115	Fr. 5,034 / £ 201	Fr. 8,544 / £ 341
Rope continuously tapering.	Kilogs. 1,594 / Lbs. 3,514	Kilogs. 2,251 / Lbs. 4,963	Kilogs. 4,783 / Lbs. 10,546	Kilogs. 2·28 / Lbs. 5·02	Kilogs. 3·21 / Lbs. 7·07	Kilogs. 6·83 / Lbs. 15·06	Fr. 2,391 / £ 96	Fr. 4,052 / £ 162	Fr. 7,653 / £ 306

Formulæ—

Round steel ropes—length m hundred metres $\begin{cases} P = 214m + 7 \cdot 2\,m^2 + \cdot 2\,m^3 \text{ tapering each 100 metres.} \\ P_1 = 3100\,(e^{\frac{7 \cdot 8 \times m}{1316}} - 1) \text{ tapering continuously.} \end{cases}$

Flat steel ropes— ,, $\begin{cases} P' = 285m + 13\,m^2 + \cdot 5\,m^3 \text{ tapering each 100 metres.} \\ P_1' = 3100\,(e^{\frac{7 \cdot 8 \times m}{10}} - 1) \text{ tapering continuously.} \end{cases}$

Flat aloes ropes— ,, $\begin{cases} P'' = 443 \cdot 7m + 31 \cdot 8m^2 + 1 \cdot 4m^3 + \cdot 08m^4 \text{ tapering each 100 metres} \\ P_1'' = 3100\,(e^{\frac{1 \times m}{\cdot 75}} - 1) \text{ tapering continuously.} \end{cases}$

For a depth of 1000 metres (1094 yards) the following are the quantities :—

	Total weight of ropes.			Average weight of ropes per running metre.			Price of ropes.		
	Round steel rope.	Flat steel rope.	Aloes rope.	Round steel rope.	Flat steel rope.	Aloes rope.	Round steel rope.	Flat steel rope.	Aloes rope.
Rope of uniform section.	Kilogs. 6,300 Lbs. 13,891	Kilogs. 23,000 Lbs. 50,715	See note.*	Kilogs. 6·30 Lbs. 13·89	Kilogs. 23· Lbs. 50·7	See note.*	Fr. 9,450 £ 378	Fr. 41,400 £ 1,656	See note.*
Rope tapering each 100 metres.	Kilogs. 3,060 Lbs. 6,747	Kilogs. 4,650 Lbs. 10,253	Kilogs. 9,817 ·Lbs. 21,646	Kilogs. 3·06 Lbs. 6·75	Kilogs. 4·65 Lbs. 10·25	Kilogs. 9·82 Lbs. 21·64	Fr. 4,590 £ 183	Fr. 8,370 £ 335	Fr. 15,697 £ 628
Rope tapering continuously.	Kilogs. 2,508 Lbs. 5,530	Kilogs. 3,563 Lbs. 7,856	Kilogs. 8,658 Lbs. 19,090	Kilogs. 2·51 Lbs. 5·53	Kilogs. 3·56 Lbs. 7·85	Kilogs. 8·66 Lbs. 19·10	Fr. 3,722 £ 150	Fr. 6,413 £ 256	Fr. 13,852 £ 554

* An aloes rope of uniform section cannot exceed 750 metres in length.

other forms of rope, induced us to make a more detailed investigation into these ropes, and the drums specially adapted for them.

(3.) The theoretical round steel rope is nearly three and a-half times as light and more economical than a rope of aloes.

The influence that the depth of the pit has upon the weight of the ropes will become more apparent by the following table.

	Rope tapering each 100 metres.		Rope tapering continuously.	
	700 metres. Kilogs.	1000 metres. Kilogs.	700 metres. Kilogs.	1000 metres. Kilogs.
Round steel ropes ...	1920 ...	3600 ...	1594 ...	2508
Flat steel ropes	2800 ...	4650 ...	2251 ...	3563
Flat aloes ropes	5340 ...	9817 ...	4783 ...	8658

We see, thus, that the 300 metres added to the end of the 700 metres increases the weight of a steel rope by more than one-half of the total weight; and nearly doubles the weight of an aloes rope.

It follows, therefore, from what has been said, that the greater the depth of the pit the more important is it to employ continuously tapering ropes, to use those made of steel, and round rather than flat ones.

We will now investigate the best method of counterbalancing the ropes and of economising fuel under the boilers of the winding engines.

METHODS OF COUNTERBALANCING ROPES.

1. ROUND STEEL ROPES.—We will assume that a round steel rope has been adopted, tapering every 100 metres, and will consider the best form of drum to counterbalance this. We must first consider what is the least radius of drum which should be employed to wind the last and stoutest coil of the rope, a coil weighing 3·348 kilogs. per running metre (8·07 lbs. per yard). This weight of rope corresponds to a section of

$$\frac{3\cdot348}{7\cdot8} \times \frac{8}{9} = 381\cdot5 \text{ square millimetres} (= 0\cdot591 \text{ square inch.})$$

We will now see what is the best way of making a rope of this section.

It has been found out by experiment that to make a thoroughly good rope all the wires must be twisted to the same extent. The wires, therefore, in each strand must be twisted round a hemp core, which will be elastic and compressible, and not round a metallic core, which would take a more direct strain and immediately break. It is for this reason that rope makers generally twist their strands with six wires round a hempen core ; and thus, also, the strands themselves should be twisted round a central core of hemp. (See Plate X. fig. 2.)

The number of wires in each strand and the number of strands in a rope are limited by practical considerations. If too many wires are twisted round the hempen core, and too many strands round the central core, the cores themselves must be very large.

And then the cores are compressed and torn by the wires which bear against each other as the rope gradually stretches, and this very soon destroys the strength of the cores, they fall into dust and are no longer of any use. M. Guillaume, of Cologne, and who was consulted by M. Kraft on this subject, was of opinion that the number of wires in each strand might readily be sixteen, but should, on no account, exceed nineteen. A rope might thus be made of seven strands with nineteen wires each, or in all 133 wires, but it would be better to reduce the number of wires in each strand to thirteen, and increase the number of strands to ten. It is a question, however, whether it would not be even better to put less wires in each strand, more strands in each rope, and to twist several ropes together round a central core of hemp (cable laid) whenever the number of wires has to exceed 100. Would it not, for instance, be better to make a large rope of six ropes twisted round a hempen core, each rope being made of six strands, and each strand of six wires, or, in all 216 wires ? All the wires, all the strands, and all the ropes would thus be twisted to the same extent, the hempen cores would be as small, and the flexibility of the rope as great as possible.

In our own case, however, we will suppose our rope to be made according to the plan usually adopted by the rope makers.

We shall see further on that the large wires generally used by them are not here detrimental to the regular working of the ropes. If we were to have a rope of six strands and eight wires in each strand, in all, forty-eight wires, that would give for the section of each wire 7·95 square millimetres (0·012 square inch) or rather more than 3 millimetres in diameter; a very stout and stiff wire.

If we were to have a rope of eight strands and eight wires in each, in all sixty-four wires, then each wire would have a diameter of 2·75 millimetres (0·1 in.)

Now, let us consider what will be the least diameter of the drum on which this rope could be wound so as not to damage it by the bending.

It will be seen that the damage done to the rope by the bending is given by the formula—

$$f = M \frac{\delta}{D}$$

Where δ = diameter of each wire.
    M   modulus of elasticity of the metal.
    D   diameter of the drum measured from the centre of the rope.
    f tension of wire produced by the bending per unit of surface.
    φ the angle at the centre of the drum subtended by the bent portion of the rope. (See Plate X. fig. 7.)

Then the arc $a\,b$ (along the neutral axis) will be neither stretched nor compressed—the arc $c\,d$ will be stretched—and $e\,f$ compressed.

The original length of all the wires will be—

$$\varphi \frac{D}{2}$$

the length of the stretched portion will be—

$$\varphi \left(\frac{D}{2} + \frac{\delta}{2} \right)$$

and, therefore, the elongation will be—

$$\varphi \frac{\delta}{2}$$

and by the elastic law (that the tension varies at the extension)—

$$\varphi \frac{\delta}{2} = \frac{f}{M} \varphi \frac{D}{2}$$

whence

$$f = M \frac{\delta}{D}$$

E

Now, the modulus of elasticity varies from 20,000 for iron wire to 27,500 for steel wire, and we can now proceed to determine the strain due to the bending which steel ropes can bear.

In the case of running ropes of iron it is usual to consider that 18 kilogs. per square millimetre (39,690 lbs. per square inch) is the limit of strain from bending and from traction which iron ropes should stand.

For steel, where the tensile strength is to that of iron, as ten to seven, according to the experiments made at Seraing, the limit should be $\frac{10}{7} \times 18 = 25 \cdot 7$ or, say, 26 kilogs. per square millimetre (57,330 lbs. per square inch) for bending and traction. As we have already determined 13 kilogs. per square millimetre as the limit of strain due to traction, there remain 13 kilogs. per square millimetre (28,660 lbs. per square inch) for the strain due to the bending.

The initial diameter of the drum will now be easily calculated from the formula—

$$f = 27,500 \times \frac{\delta}{D}$$

If $f = 12 \cdot 7$ kilogs. D = 6 metres
$f = 13$ kilogs. D = 5·8 metres.*

These diameters are rather larger than those which the English have been led to adopt in practice ; for they employ an initial diameter of drum of 4·50 metres (14 ft. 9 in.) for a rope weighing from 2·24 kilogs. to 2·48 kilogs. per running metre (from 4·9 lbs. to 5·46 lbs. per running yard) at the California Pit of the Wigan Coal and Iron Co. and at Clifton Hall.†

---

* According to Karmarsh, the absolute strength of a steel wire is given by

$$\alpha\,\delta + \beta\,\delta^2.$$

Where δ = diameter of wire, $\alpha = 21$ and $\beta = 50$ for steel; and hence the strength of a wire 2·75 millimetres in diameter will be 436 kilogs. or 73 kilogs. per square millimetre (161,000 lbs. per square inch).

It will, therefore, be seen that if we strain a wire by traction and torsion up to 26 kilogs. per square millimetre, this strain amounts to about one-third of the absolute strength.

† It is to be noted that these results are not absolutely correct, because each wire does not wind round the outside of the drum, but describes a helix as it is wound round it. It is well known that the greater the torsion of the wires, the stiffer is the rope ; in other words, the longer the pitch of the helix described by each wire the more perfect is its flexibility. For the wires, which are twisted round a hempen core, more nearly approximate in length

We shall see, further on, that these large diameters have not only the advantage of increasing the life of the ropes, but also of giving the cages a high velocity without employing a high piston-speed.

We will now calculate the final diameter and the shape which a drum must have in order to produce a constant moment of resistance.

Gerstner was the first (in his " Handbuch der Mechanik ") to give the mathematical theory of a drum to produce a constant moment of resistance. When the rope is uniform in section, the drum is conöidal, and has for a generator an S curve, instead of a straight line.

M. V. Dwelshauvers-Dery has lately produced (see the *Revue Universelle*, vol. xxxi.) the same form of drum as Gerstner, for counterbalancing ropes of a uniform section. •

These interesting investigations, however, are not applicable to a rope tapering each 100 metres, as we have assumed, nor to a continuously tapering rope. In that case we should have to adopt calculations and integrals of a very complicated character.

What we want, above all, to know here is this : is it possible for round steel ropes to work in practice under the special conditions we have assumed ? With this view we will consider now, what is the *final radius* which will give us *at the lift of the load* a moment equal to the mean moment of the useful load ; that is to say, equal to the product of this load multiplied by the mean radius of the drum, at the moment when the cages cross, and when the dead weights equalise each other, and the useful load only bears upon the mean radius.

Let $r =$ the initial radius, $R =$ the final radius, then the mean moment $=$ the useful load—

to that of the core as the torsion is less or the pitch of the helix greater. It is necessary, therefore, that the hemp core should be stout enough, and that the number of wires should be great enough, so that the torsion of the wires should be reduced to a minimum. It is thus that it appears that the values for the diameters of drums (5·8 metres and 6 metres) are the least which should be employed, and that these quantities are the more nearly correct the less the torsion of the wires in each strand, and of the strands of each rope, become. We have adopted, therefore, for greater security, 6 metres (19 ft. 8·2 in.) as the least diameter for drums.

$$\times \frac{R + r}{2} = 1600 \times \frac{R + r}{2}.$$

At the lift of the cage the moment of the resistance will $= r \times$ (total load) $+$ the weight of the rope) $- R \times$ (fixed load) $= r \times (3100 + 1964\cdot6) - R \times 1500.$[*]

If these moments be equal at " meetings" and also at the bottom, then—

$$\frac{R + r}{2} \times 1600 = r\,(3100 + 1964\cdot6) - R \times 1500$$

$$800\,R + 800\,r = 5064\cdot6\,r - 1500\,R$$

$$R\,(800 + 1500) = r\,(5064\cdot6 - 800)$$

$$R = r\,\frac{4264\cdot6}{2300} = 1\cdot85\,r$$

If these moments be equal at meetings and at the landing of the cage we get in the same way $R = 1\cdot85\,r$.

If $r = 3$ metres, $R = 5\cdot55$ metres, and the mean moment equal 6840 kilogs. (49,474·68 foot-pounds).[†]

Between these two radii the rope must wind itself on the drum in a spiral form, the drum being of a conöidal shape approaching in practice to the theoretical drum.

It is evident that these drums will have to be of enormous size when the pits are very deep. It is not, however, impossible, if they are built of wrought iron, to give them a solidity and lightness sufficient for all practical purposes. The largest drum that we met with in 1866 in England (which was·at Clifton Hall) had its largest diameter 7·50 metres (24 ft. 7·28 in.) and its form was not conöidal but half cylindrical, half conical. Between the smallest diameter, 4·50 metres (14 ft. 9·16 in.), and the largest diameter, the rope was wound along a spiral and made nine complete revolutions ; it then was wound nine times round the cylinder of 7·50 metres diameter. It wound coal from a depth of 388 metres, and had worked to as great a depth as 486

---

[*] We have seen before that the wagons, cage, and coal weigh altogether at our colliery 3100 kilogs. We have also found that a round rope 700 metres long, weighs altogether from 1930 to 1960 kilogs.

[†] It will be observed here that if we use big drums we shall be obliged to use large cylinders and powerful engines. In order to lift the load with one rope alone, the engine will have to be calculated to exert a power of 15,194 kilogrammetres (109,900·33 foot-pounds) or more than double the mean power of 6840 kilogrammetres. We shall see further on, that by using aloes ropes, we shall be able to reduce the maximum power of the engine to 9763 kilogrammetres.

metres. To do this the rope was wound four and a-half times more round the cylindrical part of the drum. I have already observed that since that time more large drums have been used for great depths, in spite of the enormous sizes of the winding engines consequent upon this.

At the Saarbrück mines they have a conöidal drum, on which the round rope is wound from an initial diameter of 4·90 metres ; it makes nineteen complete turns on the spiral, and works up to a final diameter of 8·89 metres (29 ft. 2 in.) They wind from a depth of 412·5 metres (1353·47 ft.) In the Ruhr coal basin they use wrought-iron drums whose largest diameter is as great as 10 metres (32 ft. 9·7 in.) So far as the construction is concerned it is then possible to have drums up to a diameter of 10 or 12 metres (up to 39 ft. 4·45 in.) The high price of these drums is compensated to a large extent by the economy which can be realised in the wear of the ropes, for it must be recollected that round steel ropes are twice as economical as flat steel ropes, and three times as economical as aloes ropes. For if we assume that an aloes rope worth 7653 fr. will be used every year for this purpose, it will be evident that it will be possible to save about 5000 fr. a year, and this will correspond to a capital invested of 50,000 fr. at 10 per cent. ; that is to say, to a sum much larger than the whole cost of an enormous wrought-iron drum, or even a steel drum, as well as the cost of the greater size of the engine. We must also add to this the saving of coal by means of counterbalanced drums.

As for the trouble which the enormous momentum of this drum would cause, it must not be feared. If the drums themselves are a great weight they carry only two ropes which are not heavy (4000 kilogs.), and only turn with a relatively small angular velocity ; whilst we are really turning round at a great angular velocity two ropes which weigh between them 16,000 kilogs. without considering the weight of the drums or fly-wheel. Moreover, there are at this moment in the Ruhr coalfield, drums of 10 metres in diameter which are working in practice without trouble, on account of the *vis viva* stored up in their masses.

We have now to consider what horizontal space will be necessary for the rope in the case of a scroll-drum with an initial diameter of 6 metres, and a final diameter of 11·10 metres. We must first calculate the number of turns that the rope will take on a drum such as this. The number of turns may be found by dividing the depth of the pit by the circumference, which corresponds with the mean radius of the drum.

$$\text{This mean radius} = \frac{6 + 11 \cdot 10}{4} = 4 \cdot 275 \text{ metres.}$$

$$\text{The circumference} = 26 \cdot 86 \text{ metres.}$$

$$\text{The number of turns} = \frac{700}{26 \cdot 85} = 26.$$

We will consider the horizontal distance corresponding to this number of turns. The mean diameter of the rope itself will be that of a rope weighing 2·75 kilogs. per running metre.

$$\text{The section will be } \frac{2 \cdot 75}{7 \cdot 8} \times \frac{8}{9} = 313 \text{ square millimetres.}$$

This would be equivalent to fifty-six wires each 2·75 millimetres in diameter, say eight strands of seven wires each.

The diameter of this rope is given by the formula—

$$d = \delta \left(1 + \frac{1}{\sin. \frac{\pi}{n}}\right) \left(1 + \frac{\cdot 1}{\sin. \frac{\pi}{n'}}\right)$$

Where $n =$ the number of wires in each strand
$n' =$ the number of strands.*

* This formula is frequently used in all calculations about metallic ropes, and is thus arrived at—(See Plate X. fig. 8, which represents the n wires of one strand.)

Let $\delta =$ the diameter of one wire
$d =$ 　,,　　,,　 the strand
$n =$ the number of wires in the strand.

Then it is evident that—

$$d = \frac{\delta}{2} + o b + \frac{\delta}{2} + o b$$

$$= \delta + 2 o b$$

to express $o b$ in terms of δ and n in the triangle $o c b$

$$c b \text{ or } \frac{\delta}{2} = o b \sin. \varphi$$

$$\text{or } \frac{\delta}{2} = o b \sin. \frac{\pi}{n}$$

$$\therefore o b = \frac{\frac{\delta}{2}}{\sin. \frac{\pi}{n}}$$

whence

$$d = 2\cdot75 \left(1 + \frac{1}{\sin. \frac{180}{7}}\right) \left(1 + \frac{1}{\sin. \frac{180}{8}}\right)$$

Now $\sin. \dfrac{180°}{7} = \sin. 25° \, 40' = 0\cdot433$

and $\sin. \dfrac{180°}{8} = \sin. 22° \, 30' = 0\cdot38268$

From this the result is $d = 32\cdot86$ millimetres
$= $ say 33 millimetres (1·299 in.)

and we have already found that the number of turns on the drum would be twenty-six.

We must now consider the necessary size of each groove of the spiral, so as to carry safely a rope 33 millimetres in diameter (1·299 in.)

On each turn of the spiral on the drum (see Plate X. fig. 3) a groove must evidently be made, in which the rope will rest with a certain small amount of play. This can scarcely be less than 5 millimetres (0·197 in.) A piece of plate iron at least 8 millimetres in thickness (0·315 in.) is riveted on to the drum, so as to form one of the sides of this groove. It is a good thing to give a curved slope to the edge of this plate so as to make the rope rest more securely in its groove.

According to this, the horizontal distance which must exist between the centre line of two consecutive grooves, will be on the average 46 millimetres (1·811 in.), and the vertical distance 102 millimetres (4·016 in.) Each drum containing twenty-six turns will then be 1·20 metre (3 ft. 11·2 in.) and substituting this value in the equation above—

$$d = \delta + \frac{\delta}{\sin. \frac{\pi}{n}} = \delta \left(1 + \frac{1}{\sin. \frac{\pi}{n}}\right).$$

and, further, if we were to consider each of the n' strands of the rope as one wire, we should arrive at the following expression for the diameter of the rope—

$$D = d \left(1 + \frac{1}{\sin. \frac{\pi}{n'}}\right)$$

$$= \delta \left(1 + \frac{1}{\sin. \frac{\pi}{n}}\right) \left(1 + \frac{1}{\sin. \frac{\pi}{n'}}\right)$$

broad ; and the brake which is placed between the two drums will take up a width of ·30 metre (11·81 in.), so that the whole width of the double drums will be 2·70 metres (8 ft. 10·3 in.)

The horizontal distance between the centre lines of the two head-gear pulleys is at the drawing shaft at Newville pit 1·40 metre (4 ft. 7·11 in.) ; and, therefore, the centre line of the smallest groove of the spiral, that whereon the first turn of each rope is wound is ·65 metre (25·59 in.) from the centre line of the pulley, and the centre line of the largest groove is ·55 metre (21·65 in.) on the other side of the centre line.

It has been estimated that in order that a round rope may not lie dangerously askew to the line of the pulley the farthest point of it should not lie farther horizontally from · the plane of the pulley than one-fiftieth part of the distance from the axis of the drum shaft to the axis of the pulleys. If, then, the greatest distance from the plane of the pulley be ·65 metre the highest point of the smallest groove of the drum must be 32·50 metres (106 ft. 7·55 in.) from the axis of the pulleys. Unfortunately the position of things on the pit bank at Newville would only allow of a horizontal distance of 15 metres (49 ft. 2·56 in.) between the axes of the pulleys and the drum. The difference of the two vertical heights also had to be 10 metres (32 ft. 9·7 in.) so as to give a distance of 16 metres from the bank to the axis of the pulleys and 6 metres to the axis of the drum. We adopted pulleys whose radius was 3 metres (9 ft. 10·11 in.), and, therefore, the dis- · tance from the summit of the smallest groove of the drum to the summit of the pulleys was only $\sqrt{15^2 + 10^2} = 18$ metres (59 ft. 0·67 in.) The distance, then, from the drum to the pulleys was about one-half what it should be that a round rope might run with ease. With a distance of only 18 metres the rope would inevitably cut itself against the sharp edges of the grooves at the one side of the drum, and run out of the grooves on the other side. This is just one of those troubles which arise when drums such as these are put up without a sufficiently careful design. We shall, therefore, make the case more clear yet, by indicating a method where-by the necessary distance of the drums from the pulleys can be calculated with greater accuracy still.

Let us assume the smallest groove of the drum to have a radius of 3 metres, and determine the necessary distance of this groove from the pulley, in order that the rope may never cut itself against the sharp edge of the grooves—in other words, that the rope may only just touch the edge when it leaves the drum.

We must first determine the point n (Plate X. fig. 5) where t round rope will leave the edge of the first groove of the drum.

This groove will have as its radius $o\,n$, and which is 3 metres + the height, $h\,q$ or $p\,m$, of the side of the groove against which the rope will rub. This height of the side of the groove will be equal to the vertical distance between the centres of the ropes in two consecutive turns of the groove + the height of the side above the centre of the rope. This will be $\frac{1}{25}$th of the difference between the greatest and smallest radius $(2.55) + .010 = .102 + .01 = .112.$

The distance $m\,n$ at which the rope will first leave the drum will be given by the formula

$$X^2 = 3.112^2 - 3^2$$
$$= .6845$$
$$X = .827 \text{ metres (2 ft. 8·56 in.)}$$

Now, let us investigate what actually happens in each groove. It is essential that the rope should not be at all bent on leaving the drum, in other words, that its course from the top of the drum to the pulley should be one straight line. If a is the point where the rope is tangential to the drum (see Plate X. fig. 5) c the point where it leaves the edging of the next groove, e the point where it touches the pulley, then the line $a\,c\,e$ must be a straight line. It is necessary for this that the triangle $e\,d\,a$ (where $a\,d$ is equal to the whole horizontal distance of the rope from the centre line of the pulley, and $d\,e$ the distance from the centre of the drum shaft to the centre of the pulley) should be similar to the triangle $c\,b\,a$ where $c\,b$ represents the distance X ($= .827$ metres) and $a\,b$ the play that the rope has in its groove.

In other words, it is necessary that—

$$\frac{a\,b}{b\,c} = \frac{a\,d}{d\,c}$$

We know that—

F

$$a\,d = \cdot 65 \text{ metres}$$
$$b\,c = \cdot 827 \text{ ,,}$$
$$a\,b = \cdot 0025 \text{ ,,}$$

The distance $d\,c$ will then be—

$$= \frac{a\,d \times b\,e}{a\,b}$$
$$\frac{\cdot 65 \times \cdot 827}{\cdot 0025}$$
$$= 215 \text{ metres.}$$

It is necessary, therefore, that the pulley should be distant from the drum 215 metres (705 ft. 4·7 in.) in order that the rope should not be bent against the sides of the grooves. This bending of the rope, if it were to any considerable amount, would be certain to wear it out. Now let us examine what must be the distance from the drum to the pulley in order that the rope should not tend to jump out of its groove and slip down the drum when it has passed on to the other side of the centre line of the pulley.

We must first consider the last groove of the drum which is the farthest removed from the centre line of the pulley (say ·55 metre). The distance at which the rope will touch the side of the groove, will be easily obtained as before by considering that this distance is one of the sides of the right angle of a right-angled triangle, and that the other side of the right angle is the radius (5·55 metres) of this groove of the drum, and the hypothenuse is equal to this same radius 5·55 + the height (·01 metre) of the groove above the centre of the rope, in all 5·56 metres. This distance—

$$= \sqrt{5\cdot 56^2 - 5\cdot 55^2} \,\text{,}$$
$$= \cdot 3333 \text{ metre (13·11 in.)}$$

We may calculate as before the necessary distance from the drum to the pulley in order· that the rope should not be bent by the side of the groove which is put there to prevent it from slipping off. We have—

$$\frac{2 \times 5}{\cdot 333 \times 5} = \frac{550}{y}$$
$$y = 73\cdot 30 \text{ metres (240 ft. 5·88 in.)}$$

It would, therefore, be advisable, in order to avoid accidents, to give up all idea of using scroll drums like those used in England and in Germany.

It was on this account that we were led to go back again to ·

rope rolls and flat aloes ropes. We shall give, further on, the calculations that we made for this purpose. We were, however, in writing this paper, induced to make still further researches into the question of drums, and we think now that we have arrived at a complete solution of the question.

It easily follows from what has gone before, that if instead of reducing the total play of the rope in its groove to 5 millimetres, we should increase this play to 10 millimetres (say 5 millimetres on either side of the rope) we should then want 107 metres distance between the pulley and the drum, in order that the rope should not be bent by the sides of the grooves ; and 36·60 metres in order that the rope should not fly out of the grooves.

Each drum, then, would have to be 1·326 metres instead of 1·20 metre.

It will naturally be assumed that in order to reduce as much as possible the breadth and the size of the drum, we must give only so much play to the rope as is absolutely necessary. It is thus that the groove which is exactly in the centre line of the pulley, need only have 2·5 millimetres on each side of the rope ; and as the rope is wound upon the drum and gets farther from the centre line, it will be necessary to increase the amount of play in the groove proportionately to the obliquity of the rope. In other words, the rope should, on either side of the centre line of the pulley be wound along a helix whose pitch increases in proportion to the obliquity of the rope. The sides, also, of each groove should have such a slope as to induce the rope to be wound on or off the groove without being bent at all.

We are thus naturally induced to reverse the problem and to determine the dimensions of a scroll drum which will satisfy the condition that the rope shall wind easily on and off it for a given distance from the drum to the pulleys. Let us take, for instance, our own distance, 18 metres, we must determine the amount of play which must be given to the rope on each coil successively, according to the distance from the centre line of the pulley. We will begin with the largest coil, that nearest to the brake. We know already that there is a space of ·55 metre between the edge of the brake and the centre line of the pulley, and throughout this distance the obliquity of the rope tends to make it slip down the drum.

Now, let us consider what slope must be given to the edge of the groove of the largest coil in order that the rope may not touch it.

The shortest distance between the rope and the edge of the groove (to which we must add play to the amount of 2·5 millimetres, or, in other words, the horizontal projection of the slope of which we have just been speaking (which we will call x) may be calculated by the equation—

$$\frac{x}{333} = \frac{524·7}{18000}$$

Here 333 represents the distance in millimetres which exists between the point where the outside of the rope quits the *bottom* of the groove, and the point where it quits the *edge* of the groove.

524·7 represents the horizontal distance (in millimetres) from the centre line of the rope to the centre line of the pulley-wheel (and observe that the diameter of the rope in this coil is 29·6 millimetres, and the distance from the centre line of the rope to the brake-drum is 25·3 millimetres).

18000 is the distance in millimetres from the drum to the pulleys.

From this—

$$x = 10 \text{ millimetres.}$$

Similarly for the next coil—

$$\frac{x}{330} = \frac{472}{18000}$$

whence

$$x = 8·6 \text{ millimetres}$$

and so on to the twelfth coil, where the play required will be only 2·5 millimetres, and this we shall take as the minimum.

On the other side of the centre line of the pulley (which cuts the drum at a point distant 1·6 millimetres from the edging between the eleventh and twelfth coils) the least distance (x') from the side of the rope to the edge of the groove must be deduced from the equation—

$$\frac{x'}{y} = \frac{x' + 16·6}{18000} \text{ millimetres}$$

whence

$$x' = \frac{y \times 16·6}{18000 - y}$$

Here $y =$ the distance at which the rope touches the edge

centre
.ce we

'st coil
of the

fferent
ʒ with

metres
ate X.

ɣ for a
ys and

ancing
pletely
ɔm the
. as 20

will be
ıce the
t small
n ordi-
x, with
h sides
farther
ılt con-
fficulty

ıich are
ne and
r, these
.ditions

prevent
; which
hey use

Now, l
of the
not tor
· The
the gro
millim
the slo
will ca

Her
exists
the *bot*
of the
524·
from tl
pulley-
this co
line of
1800
pulleys
Fror

Simi

whence

and so
only 2·
On t
cuts th
edging
tance (
must b·

whence

Here

of the groove, $x' + 16.6 =$ the least distance from the centre of the rope to the edge of the groove, to which distance we must add 2·5 millimetres of play.

Thus it will be found that the centre line of the first coil is distant by 18·4 millimetres from the centre line of the pulley—and hence for the thirteenth coil—

$$\frac{x'}{y} = \frac{x' + 64.6}{18000} \text{ millimetres}$$

and so forth.

We have arranged in the accompanying table the different data for determining the size of each coil, beginning with the largest coil of the drum.

We should by these data produce a drum 3·53 metres broad, including ·30 metre for the brake ring (see Plate X. figs. 6 and 7).

It is evidently a very practical design, particularly for a distance of 18 metres between the centres of the pulleys and drum.

It will be seen then that the question of counterbalancing round steel ropes by means of scroll drums is completely solved, especially if it be granted that the distance from the pulleys to the main drum-shaft is generally as much as 20 metres and often as much as 30 metres.

An objection may be urged to this drum, that it will be very difficult to build. This is true. But when once the calculations have been made for each coil, it takes but small additional trouble to build from them above building an ordinary scroll-drum. If grooves can be built upon a helix, with straight sides, it is not impossible to build them with sides which splay more and more as they get farther and farther from a certain vertical plane. Moreover, this difficult construction possesses so great advantages that its difficulty must not be considered an insuperable obstacle.

For by this means round steel ropes may be used which are three times as economical as flat aloes ropes, and one and a-half times as economical as flat steel ropes. Moreover, these round ropes may thus be placed in the very best conditions for their economical working.

By employing high edges to the grooves we can prevent the ropes from slipping down the drums, an accident which sometimes happens in England and Germany, where they use conical drums, or spiral drums with the edgings not standing up above the centre of the rope. When the rope slips down the drum the cage drops suddenly to such an extent that it generally causes either the rope or the cage-chains to break. One can scarcely imagine without a shudder what would be likely to happen if there were men in the cage.

The wear of the ropes when each coil of the rope lies in a groove which follows the natural lead of the rope itself is less than when it is wound on a plain cylindrical or conical drum, where, on account of the lateral deviation, the wires of the rope are tightly compressed one against the other, and wear each other by the friction and the bending of the rope which is induced. This destruction of the ropes on account of the lateral friction which results from one coil bearing against the other, has forced the manager of No. 1 Colliery of Levant d'Elouges to give up altogether round steel ropes in spite of the large diameter of the drum. It is necessary, to prevent the rope from being worn out too much by this winding of coil against coil, to place the pulleys sufficiently far off the drum, and that the pit should not be too deep, so as to necessitate too large a drum. A special drum is the only one which will enable the foregoing ropes to be thoroughly counterbalanced.

Conical Drum.—By winding the rope up with each coil touching the next the whole thickness of the edgings of the grooves is gained, and the lateral displacement is diminished to the same amount. This advantage, however, is gained at the cost of the counterbalancing of the ropes and their longer wear. We have seen before that in order to counterbalance the ropes for a depth of 700 metres it is necessary to employ a conical drum whose smallest radius shall be 3 metres (9 ft. 10·11 in.) and the largest 5·55 metres (18 ft. 4·5 in.)

Under these circumstances, a plain conical drum is no more applicable, first, because the slope of the surface of the drum would thus be heavier than 45°, and would cause the rope to slip down to the bottom of the drum, and secondly, because the length of the rope (700 metres) is not sufficient to enable the smallest radius, 3 metres, to pass easily to the greatest, 5·55.

Therefore, the plain conical drum will not answer for great

depths and for ropes so constructed as to necessitate a large radius of the drum.

SPIRAL TRUNCATED DRUM.—It is, therefore, necessary to employ a drum with a spiral groove. To diminish, however, the great centre of the drum, the spiral has been sometimes truncated at the point corresponding to "meetings" as we have before explained in reference to Clifton Hall Pit. The rope begins by being wound up a spiral and then is wound along an ordinary cylindrical drum, in such a manner that both the ropes are wound on to the cylinder at the moment of the "meeting" of the cages so as to produce equilibrium at that time. The ropes are now no longer counterbalanced at starting, nor for some distance from that point ; on the other hand, the rope is allowed to wind coil against coil, and the thickness of all the edgings is gained in the lateral displacement of the rope.

To make this more clear, let us take an example ; let us see how such a drum as this would work, to wind from a total depth of 700 metres (766 yards).

The useful load 1600 kilogrammes (3528 lbs.) of coal.

The cage	900	,,	(1985 ,,)	,,
Four trams	600	,,	(1323 ,,)	,,
Total	3100	,,	(6836 ,,)	,,

We will assume that the rope is to taper every 100 metres as before, that the initial diameter shall be 6 metres, and the final diameter of the spiral and of the cylindrical part of the drum 10 metres. The mean diameter of the spiral will then be 8 metres, and its mean circumference 25·12 metres. If, then, the rope takes only ten turns on the spiral, or say 251·2 metres, it will be necessary to give it rather more than fourteen turns on the cylindrical part (439·8 metres) in order to wind the whole 700 metres on the drum.

Now let us see what will be the horizontal displacement of the rope.

The fourteen turns on the cylindrical part of the drum would only occupy ·45 metres at the most, for the mean diameter of the rope throughout this portion is 32 millimetres. The breadth of the spiral portion will be ten times that of one coil, or ten times (8 millimetres + 35 millimetres) = 10

× 48 millimetres = ·48 metre. Each drum will then be ·93 metre broad ; and the total breadth of the two drums and brake will be 2·16 metres.

The horizontal distance between the two pulleys is 1·40 metre, and, therefore, it is evident that the smallest coil of the spiral is distant ·38 metre (1 ft. 2·96 in.) from the vertical plane of the pulley, and the largest coil adjoining to the cylindrical part of the drum is distant ·1 metre (3·93 in.) from the same plane.

For a distance of ·38 metre between the last coil of the spiral and the vertical plane of the pulley, it is necessary, in order that the rope should not be bent by the vertical side of the groove, that the pulley should be at a distance from the drum given by the formula—

$$\frac{·38 \times ·827}{·0025} = 126 \text{ metres.}$$

In order that the rope upon the largest coil of the drum should not tend to slip off it, the pulley must be at a distance given by the formula—

$$\frac{·333 \times ·1}{·0025} = 13·32 \text{ metres.}$$

The last coil of the rope which is wound upon the cylindrical part of the drum close to the brake ring will be distant from the centre line of the pulley by ·55 metre (1 ft. 9·65 in.) and on this account it will press against the next coil, will rub a good deal against it, and become rapidly worn out. It is evident, then, that a drum with a truncated spiral does not place the part of the rope which is wound on the cylindrical portion of the drum in good condition so far as wear is concerned ; and that when the sides of the spiral groove are left vertical the pulleys should be as far off from the drum as 104·83 metres, in order that the rope should not be destroyed by being cut against the side of the smallest groove.

The spiral portion should, therefore, have a groove with sides sloping proportionally to the lateral displacement of the rope.

Why, therefore, should we adopt a truncated spiral, and by this sacrifice to such an extent the counterbalancing of the ropes.

This is the last point that we need consider with regard to this question.

For a depth of 700 metres the moment of the moving force must be— ·

1. At the lift (1965 + 1600 + 1500) 3 − 1500 × 5 = 7695 kilogrammetres (= 55,657 foot-pounds).

2. At meetings 1600 × 5 = 8000 kilogrammetres (= 57,864 foot-pounds).

3. At the landing (1600 + 1500) 5 − (1965 + 1500) 3 = 5105 kilogrammetres (= 36,924 foot-pounds).

It will appear from this that a drum with a truncated spiral will not be a satisfactory solution of the problem of winding coal so as to preserve the ropes, and produce a constant moment of the moving force.

It will be far better to adopt a drum with a continuous spiral, with a slope in the edging of the groove proportional to the lateral displacement of the rope from the centre line of the pulley.

CYLINDRICAL DRUM.—The advantage of a cylindrical drum is that the whole rope is wound coil against coil, so that the lateral displacement of the rope from the centre line of the pulley is reduced to a minimum, and so the wear of the rope from friction is as little as possible.

Moreover, the whole rope is wound upon a drum of one (maximum) radius, so that its wear, due to the bending, is reduced as much as possible—as, also, are the number of turns of the engine in each run. It is well known that in order to draw coal from a certain depth it is important to diminish the number of turns of the engine, and not to increase too much the speed of the pistons. Since all the moving parts tend to get loosened, on account of the action of the steam, first in the one direction and then in the other.

Cylindrical drums, however, have always one great fault, which is that they do not counterbalance the ropes, and, therefore, necessitate enormously heavy engines. Let us, in order to prove this, take our own case of drawing from a depth of 700 metres, with a cylindrical drum of say 8 metres diameter.

The circumference of the drum will then be 25·12 metres, and the 700 metres of rope will correspond to twenty-eight turns of the engine.

The rope would take up a width of ·92 metre on the drum, so that if the centre line of the pulley were placed in the centre of this distance the lateral displacement of the rope would be reduced to 0·46 metre (1 ft. 6·11 in.) This would, therefore, make the rope in a good condition so far as wear is concerned, even if the distance from the drum to the pulleys were as little as 18 metres.

We will now calculate what will be the action of the moving force at the lift at meetings, and at the landing of the cage.

1. At the lift the force would be equal to (1965 + 3100) 4 − 1500 × 4 = 14,260 kilometres.

2. At "meetings" 1600 × 4 = 6400 kilometres.

3. At the landing 3100 × 4 (1965 + 1500) 4 = − 1460 kilometres.

It is evident then how much the moving force must vary, for, assuming that we can get out of a winding engine worked expansively a useful effect of 60 per cent., it is seen that we should have to exert at the beginning of the run a force of 23,766 kilogrammetres (171,899·5 foot-pounds) which would fall down to 16,667 kilogrammetres (120,552·4 foot-pounds) at "meetings" and finally become a back pressure at the end of the run of 2433 kilogrammetres (17,598 foot-pounds). With a spiral drum we should only have to develope throughout the whole run a constant force of

$$1600 \text{ kilos.} \times 4·275 \text{ metres} \times \frac{100}{60} = 11,400 \text{ kilogrammetres}$$

(82,422 foot-pounds.)

In M. Lucien Guinotte's plan of variable expansion the effect of the steam can be varied proportionately to the resistance of the load by making the grade of expansion increase accordingly. When the point of *no admission* is reached, the piston makes its own vacuum behind it, and is retarded by the pressure of the air in front. If, during this negative part of the run, steam is no more admitted to the cylinder, and, therefore, no more wasted, it is none the less true that it had been necessary to employ at the beginning of the run a relatively greater amount of power, and that it was now necessary to destroy this at the end of the run by the back pressure of the atmosphere. As soon, therefore, as an engine is made to produce a negative effect, it is no longer made to work according to a rational plan. Moreover, it is necessary that an engine should work up to its full power, in order that its useful effect should not be too much reduced, by the effect of the frictions which belong to any engine moving in vacuo.

It will not be possible to obtain in this way all the economy of fuel which will be desired, and this is the principal point to be looked after.

The conclusion is that the most satisfactory form of drum for the economy of wear of ropes and coal, is the spiral form, designed in such a way, that the pitch of the spiral and the inclination of the edgings of the grooves increase proportionally to the horizontal displacement of the rope on either side of the vertical plane of the pulley.

SIZE OF THE STEEL WIRES.—All that has gone before on the subject of round steel ropes has been based upon the use of steel wire of a diameter of 2·75 millimetres (0·108 in.) If ropes were used with much smaller wires it would be possible to reduce considerably the diameter of the drums. Now, a very usual size of wires with us is 2·2 millimetres or No. 14 gauge. If the rope were made with 7 strands it would be found that we must have twelve wires in each strand, to get a mean section of 319 square millimetres. The diameter of the rope would then be 36 millimetres. If, then, the smaller diameter of the wires allows us to use a drum, whose smallest diameter is 5 metres, and largest diameter 9·25 metres ; that is to say, if it permits us to diminish the drum to this extent, it obliges us, on the other hand, to make it broader, on account of the increase of 3 millimetres in the mean diameter of the rope. The breadth of the drum will be more than 4 metres. It will then be better to use wires pretty stout, as they have in England. It would be necessary to consider in each particular case what will be the best way to make the rope, and the best dimensions for the drum.*

* The reduction of the diameter of the wires will render it possible to counterbalance, and also not to wear out the ropes, by the use of conical drums of smaller dimensions, whenever there is sufficient distance between the engine and the pit top. Thus if a rope were made whose mean weight was 2·75 kilos. per metre, and consisting of 216 wires, each of 1·5 millimetre in diameter, viz., six ropes of six strands of six wires, the result would be a rope of rather large diameter, say 40·5 millimetres (1·594 in.) The smallest diameter of the drum might then be reduced to 3·25 metres (10 ft. 8 in.) The final diameter for counterbalancing these ropes would be 6 metres (19 ft. 8·2 in.) Moreover, if this rope, with a diameter of 40·5 millimetres, were wound upon a conical drum, with a slope of 45°, as in the case of the California pit, this rope would reach in a length of 700 metres exactly from a diameter of 3·25 metres to a diameter of 6 metres. In this way, we should attain at one time the counterbalancing and the small wear and tear of the ropes, with

It is even possible that spiral drums and steel ropes of very great strength may contain the solution of the problem of drawing coal from a depth of 1000 metres (1094 yards).

CONCLUSION.—Spiral drums, when the pitch of the helix increases with the horizontal deviation of the rope, permit the use of round steel ropes for a depth of 700 metres, and then produce an economy in the wear of the rope, and in the fuel, which is so great as to repay the very high cost of the drums themselves, and of the special engine which they necessitate, in a few years.

MEANS OF COUNTERBALANCING FLAT STEEL ROPES.

Let us assume the case of a flat steel rope which tapers each 100 metres. We can scarcely hope, in the present state of the manufacture of ropes, to get them tapering more than this. It will even be necessary to improve the manufacture to produce this result.

The universal method of counterbalancing flat ropes is by making them wind upon themselves on rope-rolls.

Let us enquire what is the smallest radius which is possible for winding the inmost turn of the rope (and, therefore, the thickest) that which weighs 5·175 kilogs. per running metre (10·4 lbs. per yard).

Suppose that the rope is made of six ropes each having thirty-six wires, or, in all 216 wires.

The maximum weight of the rope being 5·175 kilogs. per metre, will correspond to a section of $\dfrac{5·175}{7·8} \times \dfrac{8}{9} = 589·7$ square millimetres (0·913 square inch).

but small diameter of drums. In order, however, to wind forty-eight turns of each rope, and also to have a brake-drum in the centre, the whole drum must be 3·06 metres in breadth. And in order to ensure the rope winding properly on to the drum, it would be necessary that the engine should be distant from the pit at least 50 times the horizontal deviation of the rope from the centre line of the pulley. That is, 50 × ·83 = 41·5 metres, for the smallest coil of the rope. Such drums as this, then, would not do in our own case, where there is only 18 metres distance between the pit and the engine shaft. Wherever there is sufficient distance between the engine and the pit, it will generally be possible, by judiciously proportioning the diameters of the drums, and the variation of the grades of expansion, to produce a solution, not perhaps perfect, but, at any rate, practicable, of the problem of winding from a great depth.

Dividing this section by the number of wires (216) we find that the section of each wire must be 2·73 square millimetres (0·0043 square inch), and, therefore, its diameter 1·87 millimetre (0·073 in.)   This is between No. 12 and No. 13 wire gauge.

The last length of the rope would only weigh 2·983 kilogs. per running metre (6 lbs. per yard), and would only have 339·95 square millimetres (0·527 square inch) of section. This corresponds to 162 wires of 1·87 millimetre diameter or twenty-seven wires per rope.

Now, it is necessary that the diameter D_1 of the barrel of the rope rolls should be such that the strain due to the bending of the steel should not exceed 10 kilogs. per square millimetre of section.*

This must be obtained from the formula—

$$10\ e = \text{the modulus of elasticity of steel} \times \frac{\delta}{D_1}$$

Whence $D_1^1 = 5·14$ metres, if $\delta = 1·87$ millimetre, and the modulus of elasticity = 27,500.

The diameter (D_2) of the last coil of the rope may be deduced from the formula—

$$l\,d = \frac{\pi}{4}\,(D_2^2 - D_1^2) \quad \ldots \ldots \quad (1)$$

where l = the length of the rope (say 700 metres)
and d = the mean thickness of the rope.

This thickness d must be the mean of the thicknesses of

---

* According to Karmash, the absolute resistance of steel wires is given by the formula—

$$P = \alpha\,\delta + \beta\,\delta^2$$

where $\alpha = 21$
$\beta = 50$
δ = diameter of wire.

If $\delta = 1·87$ millimetre, the resistance of the wire will be $P = 214$ kilogs. (472 lbs.)   As this wire has a section of 2·75 millimetres, it is evident that each square millimetre will carry 77 kilogs. (170 lbs.) without fracture.   We have assumed that the steel wire of a flat rope ought not to be strained with more than 10 kilogs. per square millimetre of section, and we shall, therefore, assume that it must not be strained with more than 10 kilogs. by the bending. In the case of round ropes we have assumed that steel wire could be safely made to stand as much as 13 kilogs., as well by the bending as by the working load, because the wire is not weakened by the stitching, as in the case of a flat rope.   In the case of flat ropes, the total strain of the wire will then be 20 kilogs., or a quarter of its total strength.

the first and last coils of the rope. The thickness of each of these coils is the same as the diameter of the ropes of which they are made. This diameter will be deduced, as we have done before in the case of round ropes, from the formula—

$$d = \delta \left(1 + \frac{1}{\sin.\frac{\pi}{n}}\right) \left(1 + \frac{1}{\sin.\frac{\pi}{n'}}\right)$$

in which n = the number of wires in each strand.
$\qquad n'$ = number of strands in each rope.

It will thus appear that—

$$d = 15 \text{ millimetres at the top.}$$
$$= 11\cdot4 \text{ millimetres at the bottom.}$$

The mean diameter will then $= \dfrac{15 + 11\cdot4}{2} = 13\cdot2$ millimetres (0·520 in.)

From which we can calculate by means of equation (1) above, $D_2 = 6\cdot18$ metres.

We will now consider the condition in which the engine will work, or in other words how nearly the ropes will be counterbalanced with these two diameters.

At the " lift " the moment of the power will be—

$$(3100 + 2797)\ 2\cdot57 = 1500 \times 3\cdot09 = 11{,}341 \text{ kilogrammetres}$$
$$(81{,}995 \text{ foot-pounds}).$$

At the end of the run the moment of the power will be—

$$3100 \times 3\cdot09 - (2797 + 1500)\ 2\cdot57 = -1464 \text{ kilogrammetres}$$
$$(10{,}584 \text{ foot-pounds}).$$

The moment of the power will thus be negative when the cage arrives at bank, or, in other words, the engine ought to be designed so as to work rather more rationally. All negative working of engines ought certainly to be done away with.

EMPLOYMENT OF COUNTERBALANCE CHAINS.

In England, where the necessity of employing large radii for the rope-rolls, in order to work economically iron wire ropes, has been more fully appreciated than anywhere else, counterbalance chains have been used for this purpose. This counterbalance is made of large iron rings hung to the end of a chain with flat links, and working up and down a staple pit.

This chain is fastened by means of a shackle to a counterbalance-roll upon the main shaft.   The diameter of the roll, the length of the chain, and the weight of the rings are calculated so that at the "lift" the counterbalance chain is completely wound up, and counterbalances the weight of the main rope hanging in the pit.   It then is gradually let down the staple pit.   So as to counterbalance the rope as far as "meetings;" at which point the counterbalance weight should be quite at the bottom of the staple.   After this, the counterbalance chain begins immediately to wind itself up in the other direction ; so as to lift up the weight and make it counterbalance the descending rope turn by turn.   At the end of the run, the counterbalance should be completely wound up again, so as to counterbalance the weight of the descending rope.

This is the system applied in England, at Ryhope Colliery, near Newcastle, at Pendlebury Colliery, and Dankirt Colliery, near Manchester ; its advantage is that it makes the engine exert a constant power, and equal to the mean moment of the useful load.   Its disadvantage is that it necessitates the employing a special staple, which must always be kept cleaned out, and also special counterbalance chains, which are always liable to break.

We were told in England that the large rings were very apt to drop off and get injured.   There are records of Ryhope pit, of as many as four breakages of the counterbalance in one year.   It is believed that these objections can only grow worse and worse as the pits become deeper.

We have also observed that this method of counterbalancing has been abandoned at Rosebridge, where the depth is 540 metres.   By having steel cages and light trams they were enabled to use steel ropes, tapering in size from 112 millimetres to 87 millimetres, and only weighing 5·15 kilogs. per running metre.   The lightness of tapering steel ropes evidently diminishes the objection, that they cannot be counterbalanced by the difference in the radii of the rope rolls ; and as the initial diameter of the drum is 6 metres, and the final diameter 7·20 metres, it would be necessary to employ at the lift a moment of the power equal to 12,323 kilogrammetres (89,095 foot-pounds), and at the end of the run a negative moment of 1598 kilogrammetres.   The engine then would work without expansion, and very badly indeed, so far as the economy of coal is concerned. The engineman would have to alter his regulator a great deal, and make the engine work with back pressure at the end of the run. We shall show further on how much fuel may be saved by making the engine work in a more reasonable manner.

We think, therefore, that the question of employing flat wire ropes is far from being satisfactorily solved by the preceding methods. Two methods would appear to accomplish this. First, to make the ropes of the smallest possible steel wires, so that they may be wound upon a smaller diameter, and, therefore, have enough difference between the radii at the " lift " and the end of the run, to counterbalance the ropes, and avoid working by back pressure.

The second method is to give the rope as large a thickness as possible, by diminishing the number of ropes by which it is made.

We will examine the first of these methods, and consider the case of a flat steel rope with wires of 1·5 millimetre (0·058 in.) in diameter.

The small diameter of the wire is a proof of the good quality of the steel, for it is only first-rate steel which can be drawn to such a small diameter of wire. It need not be feared that the rusting, or the wear from friction, should rapidly destroy wire of No. 10 gauge. We have ourselves employed No. 10 wire to a large extent, and have proved that iron wire of this gauge is not quickly destroyed.

The first coil of the rope, which weighs 5·175 kilogrammes per metre (10·4 lbs. per yard), and has a section of 589·7 square millimetres (0·913 square inch), will be made of 334 wires each 1·5 millimetres (0·058 inch) in diameter ; or, say, fifty-six wires for each of the six ropes, i.e., seven strands of eight wires each.

The last turn of the rope, which weighs 2·982 kilogs. per metre (6 lbs. per yard) will be made of 192 wires of 1·5 millimetre (0·058in.) in diameter; or, say, thirty-two wires for each of the six ropes ; which would then be made of seven strands and four to five wires in each strand.

The smallest diameter D_1 of the drum on which the rope is wound will be given by—

$$10 \text{ kilogs.} = 27500 \times \frac{1 \cdot 5}{D_1}$$

whence $D_1 = 4 \cdot 125$ metres (13 ft. 6·4 in.)

The diameter D_2 of the drum, when the rope is completely wound up, will be given by the formula—

$$l\,d = \frac{\pi}{4}\,(D_2{}^2 - D_1{}^2) \quad . \quad . \quad . \quad . \quad . \quad . \quad (1)$$

Where $l =$ the length of the rope for a depth of 700 metres,
      $d =$ the mean thickness of the rope.
This thickness d is equal to the mean of the greatest and least thickness of the rope, and each of these may be deduced from the formula—

$$d = \delta \left(1 + \frac{1}{\sin \cdot \frac{\pi}{n}}\right) \left(1 + \frac{1}{\sin \cdot \frac{\pi}{n'}}\right)$$

in which n represents the number of wires in each strand, and n' the number of strands. From this—

$d = 11 \cdot 95\,\delta$ for the outside coil $= 17 \cdot 95$ millimetres (0·706 in.)
$d = 8 \cdot 935\,\delta$ for the inside coil $= 12 \cdot 60$ millimetres (0·495 in.)

The mean diameter will be—

15·26 millimetres (0·6 in.)

From which we deduce—

$D_2 = 5 \cdot 53$ metres (18 ft. 1·72 in.)

We will now calculate the conditions of equilibrium of the ropes.
At the lift, the moment of the moving force will be—

$(3100 + 2797)\,2 \cdot 06 - 1500 \times 2 \cdot 76 = 8007$ kilogrammetres (57,890 foot-lbs.)

At the end of the run, the moment of the moving force will be—

$3100 \times 2 \cdot 76 - (1500 + 2797)\,2 \cdot 06 = -295$ kilogrammetres (2133 foot-lbs.)

We see, then, that the difference between the moment of the power at the lift, and at the end of the run, is much less when the ropes are made of wires 1·5 millimetre in diameter.

We do not, however, avoid, as yet, the necessity of back pressure at the end of the run for a depth of 700 metres; and the greater the depth becomes the greater will this fault become.

In this case, economy of fuel will have to be sacrificed to economy in a question of smaller importance, *i.e.*, that arising

H

from substituting steel for aloes in the construction of winding ropes. On the other hand, it is to be observed that the more the initial diameter of the rope-roll diminishes with the size of the wires, the more the ring made by the rope increases in proportion ; and the greater is the fear that the rope, which has but small breadth, may some day slip down between the ring, and the horns of the rope-roll.

This objection will always be considerable in the case of tapering ropes, when the ring formed by the winding of the rope has a small initial, and a large final diameter.

The thickest end has a breadth of 107·58 millimetres (4·235 in.) and the thinnest end 75·6 millimetres (2·97 in.)

There is, then, a difference of 32 millimetres (1·26 in.) between them. It is to be observed that the fewer ropes the flat rope is made of, the greater is its thickness and the more perfect can the conditions of equilibrium become ; but, at the same time, the greater chance is there, that the rope may slip down between the ring which it forms upon the rope-rolls and the horns themselves.

CASE OF A FLAT ROPE FORMED OF FOUR ROPES.

Let us suppose the case when each rope is made of seven strands of eight wires each, fifty-six wires in all. The whole rope will thus consist of 228 wires. The section at the thickest end of the rope will be 589·7 square millimetres, in order that each of the 228 wires should have a section of 2·63 square millimetres (0·098 square inch), or a diameter of 1·83 millimètre (0·0716 in.)

The section of the thinnest end being 339·95 square millimetres (0·526 square inch), there will not be more than 152 wires in the thinnest end, say thirty-eight wires per rope, and each of these ropes will have seven strands and five or six wires in each, of 1·83 millimetre (0·07 in.) in diameter.

The thickness of the broadest end of the rope may be deduced from the formula—

$$d = \delta \left(1 + \frac{1}{\sin. \frac{180°}{8}}\right) \left(1 + \frac{1}{\sin. \frac{180°}{7}}\right)$$

$$= 1·83 \times 1195 = 21·869 \text{ millimetres } (0·863 \text{ in.})$$

The thickness at the narrowest end will be **16·6** millimetres (0·65 in.)

The breadth at the broadest end will be 87·4 millimetres (3·441 in.)

The breadth at the narrowest end will be **66·4** millimetres (2·6142 in.)

The smallest diameter of the drum will be given by the formula—

$$10 = 27500 \; \frac{1 \cdot 83}{D_1}$$

Whence $D_1 = 5$ metres.

The final diameter D_2 will be deduced from the following formula—

Where $\quad l\,d = \frac{\pi}{4}\,(D_2{}^2 - D_1{}^2)$

$l = 700,000$

$a = 19\cdot24$, the mean thickness of the rope.

$D_1 = 5$ metres.

So that $\quad D_2 = 6\cdot5$ metres (21 ft. 3·9 in.)

We will now consider in what condition of equilibrium the ropes will be.

At the "lift" the moment will be—

$5897\,R_1 - 1500\,R_2 = 5897 \times 2\cdot5 - 1500 \times 3\cdot25 = 9867$ kilogrammetres (− 71,338 foot-pounds).

At "bank" •

$3100 \times 3\cdot25 - 4297 \times 2\cdot5 = -667$ kilogrammetres (− 4822 foot-pounds).

We see, then, that the moment in this case is again negative.

We will now consider if it is possible to avoid this negative moment by employing a steel wire of 1·5 millimetre gauge (0·058 in.)

Each rope of the outside coil will have eighty-four wires, or eight strands of ten and eleven wires 1·5 millimetre diameter.

Its thickness will then be $1\cdot5 \times 15\cdot28 = 22\cdot92$ millimetres (0·902 in.)

Each rope of the lowest coil must have $\frac{192}{4}$ wires = forty-eight wires or eight strands of six wires.

Its thickness will be $1\cdot5 \times 10\cdot84 = 16\cdot26$ millimetres (0·639 in.)

The mean thickness of the rope will then be 19·59 millimetres (0·768 in.)

The smallest diameter of the rope roll will be 4·125 metres (13 ft. 6·4 in.)

The greatest diameter will be given by the formula—

$$700000 \times 19\!\cdot\!59 = 3\!\cdot\!14 \left\{ \frac{D^2}{4^2} - \left(\frac{4\cdot125}{2} \right)^2 \right\}$$

$$D^2 = 6\!\cdot\!20 \text{ metres (20 ft. 6 in.)}$$

We will now see in what condition of equilibrium the ropes will be.

At the "lift" we have—

$$5897 \times 2\!\cdot\!06 - 1500 \times 3\!\cdot\!10 = 7497 \text{ kilogrammetres (54,203 foot-pounds).}$$

At "bank"—

$$3100 \times 3\!\cdot\!10 - 4297 \times 2\!\cdot\!06 = 759 \text{ kilogrammetres (5487 foot-pounds).}$$

We see, then, that it will be possible by this means to avoid a negative moment at a depth of 700 metres.

When, however, we consider that the breadth of the outside coil is only 91·6 millimetres (3·62 in.) and that of the inmost coil 65 millimetres (2·56 in.), it is to be feared that the rope might slip down between its own coil and the horns of the rope rolls. This would be exceedingly dangerous in the case of men riding on the rope, for the rope might break from the jerk which would ensue.

We must, therefore, put out of the question the application of flat steel ropes made of four ropes, because there is no practical way of always keeping the ropes between the horns.

We doubt very much if it would be possible to invent a method of doing this : for if the rope was guided accurately by two vertical guides properly moved forward by means of screws, this would be quite certain to wear out the stitching of the flat ropes.

Moreover, the winding up of a flat rope of four ropes cannot be done so easily as one of six ropes.

For, each of the four ropes being relatively so stout tend to prevent the ropes of the next coil from being exactly superposed to these.

The ropes will, in fact, tend to arrange themselves exactly between those of the next turn. From this there results a lateral displacement of this rope, which presses some of the coil heavily against the horns. They, if they are strong enough to resist the pressure, must rapidly wear out the stitching of the rope. This circumstance must result in decreasing the life of all ropes which taper considerably in breadth.

M. Z. Velings de Châtelet, who wished to use flat metallic ropes composed of four ropes, so as to obtain a complete counterbalancing, by the differences of the radii of the rope-roll at the "lift" and at "bank," was obliged to give them up on account of the objections which have just been detailed.

If once the above theory be allowed we should, therefore, press it to its utmost limit, and reduce the number of ropes to one only ; that is to say, the case of a round rope which is wound up upon itself upon a rope roll. This is the system designed by M. Demanet, the engineer of L'Esperance collieries, at Seraing.

M. Demanet makes his round wire ropes wind in rope rolls made of metallic cheeks, which are rigidly fixed at a distance equal to the diameter of the rope. If the cheeks be made strong enough this distance can be accurately maintained.

M. Demanet starts by employing round steel ropes of *uniform* section, because they are better adapted to this form of rope-roll. .

Let us consider what will be the conditions under which they will work in our own case. A round steel rope of uniform section would weigh 3·86 kilogs. per running metre (7·779 lbs. per yard), or in all 2700 kilogs. (5953 lbs.), to draw a total load of 3100 kilogs. (6835 lbs.) from a depth of 700 metres, and according to all the practical considerations which have been detailed above.

This rope would give, according to the English tables, a diameter of 34 millimetres (1·339 in.) for ropes weighing 3 kilogs. would have a diameter of 30 millimetres, and those which weigh 3·86 kilogs. would have a diameter of—

$$\sqrt{\frac{3 \cdot 86 \times 30^2}{3}} = 34 \text{ millimetres.}$$

If this rope were made of wires 2·2 millimetres in diameter, or No. 14 gauge, which is a size very largely employed in practice, the smallest diameter of the rope-roll must be 5 metres, and the largest diameter 7·44 metres (from 16 ft. 4·854 in. to 24 ft. 4·918 in.)

The mean moment will then be—

 1600 × 3·11 metres = 4976 kilogrammetres (35,976 foot-pounds).

At the " lift " it will be—

 5800 × 2·5 − 1500 × 3·72 = 8920 kilogrammetres (64491 foot-pounds).

At " bank " it will be—

 3100 × 3·72 − (1500 + 2700) 2·50 = 1032 kilogrammetres (7461 ft.-lbs.)

We see, then, that the amount of the load will be positive at " bank " for a depth of 700 metres ; and that we can, by employing M. Guinotte's system of expansion make the power vary proportionally to the load. In this case, however, the engine will not work up to its full power.

For a depth of 800 metres (874·9 yards) it would be necessary to employ a rope weighing 4·45 kilogs. per running metre (8·96 lbs. per running yard), whose total weight will be 3560 kilogs. (7850 lbs.), and its diameter 36·5 millimetres (1·436 in.) The conditions of equilibrium will, therefore, be as follows for an initial radius of rope-roll, 2·5 metres (8 ft. 2·427 in.), and final radius 3·95 metres (12 ft. 11·5 in.)

The mean moment will be—

 1600 × 3·225 = 5160 kilogrammetres (37,306 foot-pounds).

At the " lift " it will be—

 6660 × 2·50 − 1500 × 3·95 = 10725 kilogrammetres (77,540 foot-pounds).

At " bank " it will be—

 3100 × 3·95 − (1500 + 3560) 2·50 = − 405 kilogrammetres (− 2928 ft.-lbs.)

Thus, when the cage is at " bank " the moment of the load will be negative, and the engine will be working in a manner far from economical, with the load compelled to come to bank without steam by reason of its being over counter-balanced.

M. Demanet, however, assumes, with the English rope makers, that a steel rope weighing 1 kilog. to the metre can carry a working load of 2000 kilogs ; and this assumption puts his system in a better light. A rope weighing 3 kilogs. to the running metre, would weigh for a depth of 800 metres (875 yards) 2400 kilogs. (5292 lbs.), and would have a diameter of 30 millimetres (1·181 in.) If the least diameter of the rope-roll be 5 metres, the largest diameter will be 7·40 metres (24 ft. 3·344 in.)

The mean moment of the load will be—

1500 × 3·10 = 4650 kilogrammetres (33,619 foot-pounds).

The moment at the "lift" will be—

(3100 + 2400) 2·50 — 1500 × 3·70 = 8200 kilogrammetres (59,286 ft.-lbs.)

The moment at "bank" will be—

3100 × 3·70 — (1500 + 2400) 2·50 = 1720 kilogrammetres
(12,435 foot-pounds).

There will, therefore, be a *positive* moment when the cage is at "bank." We do not, however, think that it is prudent, to allow the load on a steel rope to exceed 13 kilogs. per square millimetre of section (0·0286 lbs. per square inch).

This system enables far better conditions of equilibrium to be obtained than are possible with flat ropes. It also employs round ropes which are much more economical in use than flat ropes. It possesses the advantage over systems with drums that it does not necessitate such large diameters as they do.

On the other hand, the following objections have been urged against it.

1. It is to be feared that a rope, where each coil bears on the neighbouring coils along one line only, will quickly become damaged. In order to counteract this tendency, M. Demanet proposes to manufacture the rope of coarser wires, which will neither wear nor break so easily as fine ones ; and also to twist them very sharply, so as to diminish the flattening of the rope between the cheeks of the rope-roll. Ropes made on this principle—of course, wires sharply twisted—are possible only if the radii of the rolls be sufficiently large. In this case, however, the conditions of equilibrium become worse. It remains to be seen how such ropes would behave in practice.

With a spiral drum the economical wear of the rope is insured, for the rope lies in a groove, between whose walls it lies evenly, without being even disturbed, owing to the oblique direction to the pulley ; moreover, it receives no pressure from the coils outside, tending to flatten it.

2. It becomes more difficult to employ ropes which are tapering in section, and this is quite indispensable when working from great depths. It is necessary that the coils, in spite of their tapering, shall wind on the top of each other, and that without catching against each other, or against the horns. It is for this end that M. Demanet has proposed to

make the hempen cores of the ropes as stout as possible, and also to entirely enclose the rope in gutta percha or in hemp of a certain thickness, so as to equalise the section of the rope. This coating will also prevent the rope from being worn out, because the whole work of the winding is concentrated on one series of edges. The only question is, will not this coating, whether made of hemp or of gutta percha, be itself cut to pieces, because the iron wire will be pressing against it with all the power due to the weight hanging in the pit? It will remain, therefore, to employ the above coating, and then to try if an ordinary steel rope which tapers each 100 metres and decreases from 32·8 to 26·7 millimetres, or only 6 millimetres (0·236 in.) in all, will not wind properly between two sets of horns which are only distant 34 millimetres (1·339 in.) from each other. In any case, since round ropes have no stitching, they will always work better than M. Velings' ropes, which are made of four ropes. The system would seem to have several important advantages. It would be interesting to see it put in actual practice, and M. Demanet is going to try it on a small scale at an ironstone pit.

The conclusion that we arrive at is, that in order to ensure a proper degree of economy in the case of flat steel ropes, wound upon rope-rolls, it is necessary to sacrifice the counterbalancing of the ropes, and incur a proportionally larger cost in fuel; which cost will go a long way towards destroying the great economy of steel ropes over those made of aloes, as we shall see further on. We, therefore, conclude that we must give up using metallic ropes wound upon rope-rolls in working from great depths.

METHODS OF COUNTERBALANCING FLAT ALOES ROPES.

Let us consider the case of a flat aloes rope, tapering each 100 metres, and working in similar circumstances to the foregoing steel ropes. It is well known that makers can spin ropes tapering every 50 metres, and even every 25 metres, without splicing them. It would even be possible to make the tapering go farther still, by dropping out not several threads every 25 metres, but each thread successively. The

manufacture would become more expensive, because it would
be necessary to stop the machinery to cut out each thread.
Up to the present time they have not been able to make
ropes with economy tapering oftener than every 25 metres ;
for they do not gain sufficiently in the weight to counterbalance the extra cost.

A rope of aloes made of six ropes, and tapering every 100
metres, and calculated according to the table given by
Messrs. Vertongen and Göens, of Termonde, will have the
following dimensions :—

For a depth of 700 metres (766 yards).

Section.	Weight of each section.		Breadth and thickness.	
	Kilogs.	Lbs.	Millimetres.	Inches.
1	467	1030	145 × 32	5·709 × 1·260
2	566	1248	160 × 35	6·299 × 1·378
3	639	1409	170 × 37·6	6·693 × 1·480
4	720	1587	180 × 40	7·087 × 1·575
5	844	1861	195 × 43	7·677 × 1·693
6	979	2159	210 × 46·6	8·268 × 1·834
7	1125	2481	225 × 50	8·858 × 1·968

Totals... 5340 ...11,775

For a depth of 800 metres (875 yards).

Section.	Weight of each section.		Breadth and thickness.	
	Kilogs.	Lbs.	Millimetres.	Inches.
1	467	1030	145 × 32	5·709 × 1·260
2	566	1248	160 × 35	6·299 × 1·378
3	639	1409	170 × 37·6	6·693 × 1·480
4	720	1587	180 × 40	7·087 × 1·575
5	844	1861	195 × 43	7·677 × 1·693
6	979	2159	210 × 46·6	8·268 × 1·834
7	1125	2481	225 × 50	8·858 × 1·968
8	1323	2917	245 × 54	9·646 × 2·126

Totals... 6663 ...14,692

It will be observed that, according to the list of Messrs.
Vertongen and Co., we ought to give to the different sections
weights which differ by several kilogs. from those which were

calculated before, and based upon the assumption that aloes rope which weighs 1 kilog. per running metre will carry a working load of 750 kilogs. The reason of this is that the weights have to be adjusted to the breadths and thicknesses which the manufacture of the rope requires. Moreover, the final result is almost the same, for we have deduced as the total weight 5340·7, and according to the list the united weights of these sections would be 5340 kilogs.

An aloes rope made of eight ropes and tapering each hundred metres would have, according to M. Vertongen's list, the following dimensions.

For a depth of 700 metres (766 yards).

Section.	Weight of each section.		Breadth and thickness.	
	Kilogs.	Lbs.	Millimetres.	Inches.
1 ...	479 ...	1056 ...	170 × 28 ...	6·693 × 1·102
2 ...	570 ...	1257 ...	185 × 31 ...	7·284 × 1·220
3 ...	632 ...	1394 ...	195 × 32 ...	7·677 × 1·260
4 ...	735 ...	1621 ...	210 × 34 ...	8·268 × 1·339
5 ...	844 ...	1861 ...	225 × 38 ...	8·858 × 1·496
·6 ...	980 ...	2161 ...	245 × 40 ...	9·646 × 1·575
7 ...	1134 ...	2500 ...	270 × 42 ...	10·630 × 1·654
Totals ...	5374 ...	11,850		

For a depth of 800 metres (875 yards).

Section.	Weight of each section.		Breadth and thickness.	
	Kilogs.	Lbs.	Millimetres.	Inches.
1 ...	479 ...	1055 ...	170 × 28 ...	6·693 × 1·102
2 ...	570 ...	1257 ...	185 × 31 ...	7·284 × 1·220
3 ...	632 ...	1394 ...	195 × 32 ...	7·677 × 1·260
4 ...	735 ...	1621 ...	210 × 34 ...	8·268 × 1·339
5 ...	844 ...	1861 ...	225 × 38 ...	8·858 × 1·496
6	980 ...	2161 ...	245 × 40 ...	9·646 × 1·575
7 ...	1134 ...	2500 ...	270 × 42 ...	10·630 × 1·654
8 ...	1305 ...	2878 ...	290 × 45 ...	11·417 × 1·772
Totals ...	6679 ...	14,727		

The weight of the theoretical rope of 700 metres in length

is 4783 kilogs. (10,547 lbs.) ; and of one 800 metres in length, 5906 kilogs. (13,023 lbs.)

Let us now calculate the radius of roll which will counterbalance aloes ropes tapering each 100 metres, and of a length of 700 and 800 metres.

We will first take a rope made of six ropes and 700 metres long.

The initial radius R_1 and the final radius R_2 must satisfy the following equation :—

$$l\,d = \pi\,(R_2{}^2 - R_1{}^2) \quad \ldots \quad \ldots \quad \text{(1)}$$

Where l = length of rope

and d = mean thickness of rope.

Whence $700 \times 4\cdot06 = 3\cdot14\,(R_2{}^2 - R_1{}^2)$

or $R_2{}^2 - R_1{}^2 = 9\cdot05$

These same radii must also satisfy the equation of equilibrium of the moments.

$$(5340 + 3100)\,R_1 - 1500\,R_2 = 3100\,R_2 - (5340 + 1500)\,R_1 \quad \ldots \quad \text{(2)}$$

$$\text{Whence } R_2 = \frac{15280}{4600}\,R_1$$

$$= 3\cdot32\,R_1.$$

Substituting this value for R_2 in equation (1) we have—

$R_1 = 0\cdot96$ metre (3 ft. 1·8 in.)

$R_2 = 3\cdot19$ metres (10 ft. 5·6 in.)

If the rope be made of six ropes, and be 800 metres in length, we find that—

$R_1 = 0\cdot872$ metre (2 ft. 10·3 in.)

$R_2 = 3\cdot398$ metres (11 ft. 1·8 in.)

If the rope be made of eight ropes, and be 700 metres in length we have—

$R_1 = 0\cdot85$ metre (2 ft. 9·46 in.)

$R_2 = 2\cdot836$ metres (9 ft. 3·65 ins.)

and if it be 800 metres in length—

$R_1 = 0\cdot805$ metre (2 ft. 7·69 in.)

$R_2 = 3\cdot143$ metres (10 ft. 3·74 in.)

If the foregoing values do not produce as complete an equilibrium in the ropes as those of Combes, or those worked out by our Professor Trasenster in his lectures on practical mining,* they are sufficient, at any rate, to prove that it is

---

* M. Dwelshauvers-Dery has also produced in the *Revue Universelle des Mines*, a method of counterbalancing ropes on a rope-roll. He shows that the moment of the load can be made perfectly constant, by employing a rope of a constant thickness, but of a section which varies uniformly, and

possible to arrange aloes ropes so as to be counterbalanced with a depth of 700 metres, and that without so far reducing the initial diameter of the rope-rolls as to interfere with their lasting properly.

We have, in the Liége district, aloes ropes which always last out their warranty of two years' duration, working upon a rope-roll whose diameter is 2 metres (6 ft. 6·74 in.)

The rope made of six ropes is counterbalanced for a depth of 700 metres, with a diameter of rope-roll of 1·92 metre (6 ft. 3·592 in.), or as nearly two metres as may be. This rope will be better than that made of eight ropes, which require an initial diameter of rope-roll of 1·70 metre (5 ft. 6·93 in.)

If the rope be tapered every 50 metres, or even 25 metres, instead of every 100 metres, the perfect counterbalance with a diameter of rope-roll of 2 metres would ensue. It is scarcely prudent to make the rope-roll smaller than this if the ropes are to work as economically as possible ; and, therefore, in the case of depths which are above 700 metres it will be advisable to sacrifice the perfect counterbalancing of the ropes, in order to ensure their working economically. The adoption of a variable expansion will altogether obviate any evil result from this cause.

It has, therefore, been proved that, by using aloes ropes made of six ropes and wound upon rope-rolls, it is possible to produce a perfect counterbalance for a depth of 700 metres. It is evident, therefore, that the work which the engines have to do has been equalised for the whole run, and, therefore, that a fixed grade of expansion may be employed.*

combining it with a size of rope-roll correctly proportioned to it. For a depth of 900 metres the moment of the load will become practically constant if the rope be 4 centimetres (1·575 in.) in thickness, and the rope-roll have an initial radius of 1·326 metre (4 ft. 4·206 in.) We must, however, observe that it is hardly possible to make the ropes taper in breadth without, at the same time, tapering them in thickness; and, moreover, that the section of the rope should not vary *uniformly* with the depth, but according to a more rapid progression than this.

* The above statement would not seem to be exactly correct. When an engine has to start a certain fixed load from rest, and, after moving it with a certain final velocity, retard it gradually until it is at rest again (and this is really the case which the author is considering), it is clear that the accelera-

For depths above 700 metres an expansion, varied but slightly, will produce a variation in the power of the engines proportional to the variation in the load.   The engines can, therefore, be set to work very nearly in their best conditions. Aloes ropes, therefore, allow of a thoroughly economical employment of the steam, and, therefore, of coal.   As for the economy of the ropes themselves, this is insured if the diameter of the rope-roll be as great as 2 metres (6 ft. 6·742 in.) Their large breadth of bearing surface makes it out of the question for them to slip down between their own coils and the horns.   We have in use ropes of 800 metres in length, and tapering every 50 metres, which wind and unwind with perfect safety.

In the case of tapering ropes without splices it is not possible to alter the " lift " as in the case of ordinary ropes.* It is essential, therefore, to have the rope longer than is absolutely necessary, and to alter the point of the rope, which receives the greatest strain from the " lift," every month, or at least pretty often, by cutting a piece off the end, and making the whole rope move outwards down the pit by that amount. If this be done it is safe according to M. Vertongen to load the part of the rope at the " lift " with 0·75 kilogs. per square millimetre of section (1654 lbs. per square inch) in the same way as the middle of the rope.   If this be done it will be necessary to reduce the size of the centre bars of the rope-rolls to suit the increased length of rope.   In our own case, it would be necessary to reduce it to 1·20 metre in diameter (3 ft. 11·245 in.)   The working part of the rope will still wind itself on and off the rope-roll from an initial diameter of 2 metres.   The repairs to a rope of aloes such as this, are easily made by using a machine to compress the piece of rope spliced in, and to reduce it to the proper size to suit the

tion at the beginning must be produced by a greater effort in the engines, and the retardation at the end by a less effort.   The amount of work done *per revolution* will be the same if the diameter of the rope-roll does not vary, but not the amount *per minute*.   Moreover, all engines, to some small extent, require a larger admission of steam, or, in other words, a less grade of expansion, as their speed increases.—*Translator.*

* It would appear from this that the practice in Belgium is to cut and re-splice the thickest section of the rope, and so alter the position of the " lift " by turning it end for end.   I need scarcely add that such a practice would be utterly repugnant to all English instincts.—*Translator.*

various sizes of rope in different parts. An aloes rope has this great advantage over all others, that it enables a good rope maker to detect the bad places, and catch them before the rope breaks. This advantage is sacrificed in the case of metallic ropes. In their case, in order to be safe, it is necessary to take off the rope when it has run its regular time, a time determined by careful experiments. Thus, in England, metallic ropes are generally replaced after fifteen or eighteen months.*

Round steel ropes, however, must eventually supplant flat aloes ropes. They are nearly three times as light and as cheap as the latter ; and our form of spiral drum, combined with them, makes it possible to obtain, together with this great economy in ropes, as great an economy of fuel as with rope-rolls.†

WINDING ENGINES.

We will now proceed to investigate the power and the proportions of the winding engines.

We shall then have to enquire what class of engines will enable us to develope this power as economically and simply as possible.

Lastly, we shall examine into the arrangements which will become necessary to tie the framing of the engines to the head gear.

1. THE POWER AND PROPORTIONS OF WINDING ENGINES.—A pair of winding engines must be so designed that the two

---

* I think that English engineers will hardly endorse such an opinion as this. A careful examination of wire ropes *day by day*, and a careful record of any broken wires, the lubrication of them with some tolerably thin oil, which permeates the rope, and does not disguise its surface, and the cutting periodically of the end, and examination of the part cut off, so as to see the condition of the wires themselves, afford a capital test of the state of the rope, independently of the time it has been at work. The "life" of a rope depends entirely on the work which it has to do compared with its strength, and, at any rate, varies from six months to three years.—*Translator.*

† We much regret that we did not discover the form of spiral drum which is applicable to a great depth of pit until after we had laid the foundations of our new engines, and all was too far advanced to enable us to retrace our steps. We should have adopted, instead of flat aloes ropes, the most tapering form of steel ropes which we could get, and have wound them upon the drums described above.

cylinders together are powerful enough to lift the total load when one rope alone is fitted.

The load of coal will be	1600 kilogs.	= 3528 lbs.	
The weight of cage and trams	1500 ,,	= 3307 ,,	
	—— 3100 kilogs.	= —— 6835 lbs.	
The weight of a tapering aloes rope	6663 ,,	= 14,692 lbs.	
The effective pressure in the boilers, 4 atmospheres		= 60 lbs.	
The initial pressure in the cylinders, $3\frac{1}{2}$ ·,,		= $52\frac{1}{2}$,,	
The mean speed of winding, 10 metres per second		= 1969 ft. per min.	

The whole depth of 800 metres will be gone over in 80 seconds.

The number of turns in one run will be obtained by dividing the difference between the initial and final radii of the drum by the mean thickness of the rope.

This will be—

$$\frac{3\cdot43 - 1}{0\cdot04227} = 57\cdot4 \text{ turns.}$$

and each turn will be performed in—

$$\frac{80}{\cdot 57\cdot4} = 1\cdot41 \text{ seconds.}$$

If we assume a stroke of 1·20 metres (3 ft. 11·245 in.) we shall have for the piston speed—

$$\frac{2\cdot40}{1\cdot41} = 1\cdot70 \text{ metre per second (335 ft. per minute).}$$

The necessary diameters of the cylinders will be given by the condition of lifting the whole load with one rope.

The moment of the load will be obtained by multiplying the total load added to the weight of the rope by the radius of the drum at the lift ; which radius must not be smaller than 1 metre in order to ensure the economical working of rope.

The moment of the resistance will then be—

$$9763 \times 1 = 9763 \text{ kilogrammetres (70,586 foot-pounds).}$$

We will now calculate the moment of the power. For this we must take into account the obliquity of the cranks. The total mean moment of the steam pressure in both pistons is equal to the surface of one piston, multiplied by the pressure of steam, by the radius of the crank, and by the co-efficient 1·273. For the mean moment of the two cylinders is equal to the power acting upon one piston multiplied by the radius of the crank and this same quantity, 1·273—

For if P = the power applied to one piston
  r = the radius of the crank
  Q = the total load
  R = the mean radius of the rope-roll
  S = the surface of the piston

then the work done by the power during one complete revolution will equal the work done upon the load.

The work done by the power is for one cylinder—

$$4 \times P\,r, \text{ and for both cylinders } 2 \times 4\,P\,r.$$

The work done upon the load through one complete revolution will be—

$$2\,\pi\,R \times Q$$

therefore $2{\cdot}4\,r\,P = 2\,\pi\,R\,Q$

whence $R\,Q = \dfrac{8\,r\,P}{2\,\pi} = \dfrac{4\,P\,r}{\pi} = 1{\cdot}273\,P\,r.$

The mean moment, then, for two cylinders = $1{\cdot}273\,P\,r.$

To resume—we can, in no case, count upon more than 65 per cent. of the pressure of the steam as transmitted to the main shaft—

and since $\quad$ P = S × pressure of steam
  $= S \times 3{\cdot}5 \times 1{\cdot}033$

therefore $9763 = 1{\cdot}273 \times P \times r \times 0{\cdot}65$
  $= 1{\cdot}273 \times 3{\cdot}5 \times 1{\cdot}033 \times S \times 0{\cdot}60 \times 0{\cdot}65$

whence $\quad S = \dfrac{9763}{1{\cdot}7952} = {\cdot}5493$ square metres.

The diameter then of each cylinder will be 0·84 metre (33·09 in.) If it be noted that in regular working the moment of the power will be 4618 kilogrammetres (33,388 foot-pounds) it will be seen that the engine would be more than twice as powerful as necessary, with cylinders 0·84 metre in diameter. We decided, however, to increase the diameter to 0·9 metre (35·434 in.), and to make the engine work expansively for its ordinary work.

It will be invariably necessary to "change" with the full pressure of steam ; for it often happens that then one piston is on the dead point, and, therefore, the other one alone must turn the engine round. Moreover, the work done by the engines when "changing" is more than ordinary ; for it is the custom to allow the cage at the bottom to lie upon the props whilst the cage which is at "bank" is lifted up off the props in order to "change" it.

2. GENERAL DESIGN OF ENGINES.—Our winding engine

would have to raise 1600 kilogs. (3528 lbs.) of coal at a mean speed of 10 metres per second (1969 ft. per minute). It would have to exert a useful effect in the pit of 213 horse-power, or a theoretical power of 327 horse-power (if we calculate on realising 65 per cent. as above). It is evident, therefore, that it will be of enormous importance to employ a class of engine which will consume as little coal as possible, without serious complication.

It is a pretty general proportion to calculate that a colliery which produces 200,000 tons of coal per annum will use as colliery consumption about 5 per cent. of that quantity, or say 10,000 tons. This quantity, taken at a mean price of 8 fr. per ton (reckoning the actual cost to the colliery alone), would make the cost of the colliery consumption alone 80,000 fr. (£3200). If the coal be taken at its selling price the amount would come up to 200,000 fr. (£8000).

This consumption of coal can be reduced by more than one half, by having a high-pressure engine, working expansively, and also condensing, if there is sufficient water supply.

We will take each of these points in detail.

By condensing the steam, it is possible to reduce the back pressure to an amount represented by a column of mercury of the height of 0·16 metre (6·299 in.) instead of the atmospheric pressure represented by a column of mercury 0·76 metre (29·922 in.) in height. It is possible, therefore, to gain 0·6 metre (23·622 in.) of vacuum, or $\frac{1}{44}$th of the effective pressure of three and a-half atmospheres, which correspond to a column of mercury 2·66 metres (8 ft. 8·727 in.) in height.

The consumption of coal in a winding engine of 300 horse-power working for eighteen hours, will be 17,820 kilogs. (39,293 lbs.) per day, for it will take 3 kilogs. (6·6 lbs.) of coal per horse-power per hour. At the cost of 11 fr. per ton, this will be equivalent to 178 fr. for coal per day. By working the engine condensively, we may then save $\frac{178}{44}$ or 40 fr. per day.

Unfortunately, a condensing engine necessitates a large water supply. According to Armengaud, it takes 300 kilogs. (661·5 lbs.) of water per horse-power per hour to condense the steam with ordinary conditions of pressure and expansion. Our winding engine would then want, allowing for stoppages,

K

70 cubic metres of water per hour; and in dry times our water supply, which was used for the boilers as well, would only give us 15 cubic metres per hour. In order, therefore, to get sufficient water supply, it would have been necessary to have brought the water from the river Sambre by a conduit 1600 metres long, and to pump it up into this by means of a special steam engine.

If we were to add to the daily cost of 10 fr., which corresponds to this apparatus, a further daily cost in coal and wages of 30·5 fr., we should have altogether a cost of 40·5 fr. per day to charge to the head of condensation.

This cost would counterbalance the economy of condensing our steam, and we were naturally obliged to give it up on account of the mechanical complications it would necessitate.

In all cases, condensation should be used wherever an abundant water supply can be got at a reasonable cost. It is especially useful with an engine working at a medium pressure. It would then be advisable to employ an air pump worked by an independent engine, and not from the winding engine; so as not to complicate the latter, and to have always a vacuum to start it. It might thus be possible to have one condensing apparatus for all the engines of the colliery.

EMPLOYMENT OF HIGH PRESSURES.

The advantages of condensation diminish as the pressure of the steam increases. If, instead of an effective pressure of three and a-half atmospheres, we had a pressure of four and a-half atmospheres, the effect of condensation, instead of producing an economy of coal of $\frac{1}{4\cdot4}$th of the total consumption, would only produce an economy of $\frac{1}{5\cdot7}$th of the same quantity. As simplicity is a necessary condition of all winding machinery, it is evident that condensation must be given up in the case of all engines working at a high pressure. High pressure again enables us to reduce the dimensions of the pistons and of the cylinders, if the steam is not used expansively or to increase the grade of expansion at which the engine works. In this case it would be better to use engines working at the effective pressure of six or seven atmospheres, in any new pits that may be set out. In the case of locomotives they have already employed a pressure of ten atmospheres. At the No. 1 pit of the colliery of the Pays de Liége, where the boilers were tested for an effective pressure of four atmospheres, we adopted the opinion that has just been given. Moreover, with the consent of the Office of Mines, we had our boilers passed for an effective pressure of five atmospheres, after having had them proved up to ten atmospheres, and having obtained from a boiler maker a declaration that our boilers would work at the pressure of five atmospheres without any danger. ·

WORKING WITH EXPANSION.

Besides its regular advantages, expansion has these special advantages when applied to winding engines.

1. It is well known that winding engines are designed to draw the whole weight with one rope alone.

They must, therefore, be two or three times as strong as is necessary for their regular work.   It is obvious that it would be better to use the steam expansively, rather than to throttle it by means of the regulator.

2. In those numerous cases, where the moment of the resistance is not constant as is always the case with flat wire ropes, it is necessary to make the power developed during each revolution vary proportionally to the resistance, by means of the different grades of expansion, and not by the barbarous practice of throttling the steam.

3. Expansion diminishes the throttling of the steam in its admission to the cylinder and its back pressure during the exhaust stroke, which result from too great piston-speed, from too little lead on the exhaust side of the valve, and from too small steam and exhaust ports, such as are usually met with in winding engines.

Under these circumstances the economy which may be realised by the application of expansion will amount to 60 per cent.*

* M. Scohy has given an account in the "Bulletin" of the old pupils of the Hainault School of Miners (April, 1870) of one of his winding engines where the area of the steam ports was only $\frac{1}{40}$th of the area of the piston, and that of the exhaust port only $\frac{1}{30}$th of the same area, where the lead on the steam side was zero, and that on the exhaust side 10 millimetres out of a total breadth of 40 millimetres.   The results were as follow :—

At the 3rd revolution, with a piston speed of 1 metre, the back-pressure was 50 kilogs. per square centimetre.

GRADE OF EXPANSION TO BE ADOPTED.

The following is extracted from a paper upon this subject by M. Kraft :—

"The dimensions of the steam cylinder must first be determined by the condition that it must be able to lift the total load of one side of the pit, without any counterbalance, but with no expansion. It will then be too large for the regular work of the engine. The question then arises whether it is better to work with as high a pressure as is possible in the boilers, with a high grade of expansion, even so as to reduce the final pressure to below that of the atmosphere ; or to work altogether with a low boiler pressure and low grade of expansion.

"Now let—

T = the work to be done at each stroke of the piston, which will be constant.

S = the area of piston (constant also).

l = length of stroke (constant).

P = initial pressure in cylinder. This will vary according to the grade of expansion which is to be employed.

At the 45th revolution, with a piston speed of 2 metres, the back-pressure was 50 kilogs. per square centimetre.

At the 75th revolution, with a piston speed of 2·50 metres, the back-pressure was 70 kilogs. per square centimetre.

M. Scohy reduced this back-pressure a little by increasing the lead on the exhaust side ; but even then the back-pressure was far too heavy. For in the case of engines which move with a high piston speed, the area of the steam ports should be $\frac{1}{20}$th of the area of the piston, and that of the exhaust port $\frac{1}{15}$th of the same area, and the lead on the exhaust side $\frac{1}{3}$rd of the breadth of the port. These ports must also be doubled, so as to present a good broad passage for the steam. Otherwise, if there be a large piston speed, the following are the results :—(1.) The steam is wire-drawn, owing to the large piston area as compared with the steam-ports ; this also diminishes the working pressure. (2.) Back-pressure in the exhaust, owing to the large masses of steam which are suddenly let out through contracted and tortuous passages. In such a case, a high grade of expansion (1) reduces the amount of steam used in each stroke, and, therefore, also the amount of wire drawing and consequent reduction of pressure ; (2) it diminishes the pressure of steam before it is exhausted, and, therefore, almost destroys all back-pressure. It is found by experiment that expansive working increases the initial pressure and decreases the back-pressure. It was concluded that the expansion gear remedied the faulty construction of the winding engine, and the too large size of the piston with reference to the work to be done, and that the result was an economy of 60 per cent.

m = grade of expansion, *i.e.*, the ratio of the whole stroke to that portion of it during which the steam is admitted. This is a variable quantity.

p = pressure of the atmosphere.

Δ = density of the steam when admitted to the cylinder, or the weight of a cubic metre of this steam. Approximately $\Delta = a\,P$, a being a constant.

γ = weight of steam used per stroke of piston.

"Then the *mean pressure* of the steam in the cylinder during the half stroke is given by the formula—

$$P\frac{1 + \log m}{m} \qquad \dots \dots \dots \quad (1)$$

"The work done during one half stroke is—

$$T = S\,l\,P\left(\frac{1 + \log m}{m}\right) - S\,p\,l \quad \dots \dots \quad (2)$$

"The amount of steam used is—

$$\gamma = \frac{S\,l}{m}\,a\,P \qquad \dots \dots \dots \quad *(3)$$

---

* These expressions may be arrived at as follows :—

1. The mean pressure is most easily deduced from the theoretical indicator diagram of the half stroke.

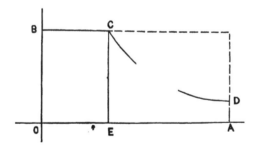

Let $O\,A = l$ the length of the stroke.

$O\,B = P$ the initial pressure of steam.

$O\,E = \dfrac{l}{m}$ the point of the cut-off.

Then the area of the diagram $O\,B\,C\,D\,A = l \times$ mean pressure.

So that the *mean pressure* required $= \dfrac{\text{area } O\,B\,C\,D\,A}{l}$.

To find this area it is to be noted that the curve $C\,D$ approximates in form to the arc of a rectangular hyperbola of which $O\,A$, $O\,B$ are asymptotes.

Now the equation to this curve (referred to its asymptotes as axes) is—

$$x\,y = c^2$$

"From equation (2) we have—

$$P = \left(\frac{T}{S\,l} + p\right) \frac{m}{1 + \log. \, m}$$

"Substituting this value in (3) we have—

$$\gamma = \alpha \, \frac{(T + p\,S\,l)}{1 + \log. \, m} \quad . \quad . \quad . \quad . \quad . \quad . \quad . \quad . \quad (4)$$

"Therefore as m increases γ diminishes, and it is, therefore, economical to work with a high initial pressure, and high grade of expansion. This result is, of course, only approxi-

---

and the area $\mathrm{E\,C\,A\,D} = \int_{O\,E}^{O\,A} y\,d\,x = \int_{\frac{l}{m}}^{l} y\,d\,x$

$$= \int_{\frac{l}{m}}^{l} \frac{c^2\,d\,x}{x} \text{ by the equation to the curve}$$

$$= c^2 \left(\log. \, l - \log. \, \frac{l}{m} \right)$$

$$= c^2 \log. \, \frac{l.m}{l} = c^2 \log. \, m.$$

Again, because c is a point on the curve, and

$$\mathrm{C\,E} = P \text{ and } \mathrm{O\,E} = \frac{l}{m}$$

$$\therefore \, P. \, \frac{l}{m} = c^2 = \text{area } \mathrm{O\,B\,C\,E}$$

so that the area $\mathrm{O\,B\,C\,D\,A} = \mathrm{O\,B\,C\,E} + \mathrm{E\,C\,D\,A}$

$$= \frac{P.l}{m} + \frac{P.l}{m} \log. \, m = P.l. \, \frac{1 + \log. \, m}{m}$$

and the area $\dfrac{\mathrm{O\,B\,C\,D\,A}}{l} = P. \, \dfrac{1 + \log. \, m}{m}$

2. The work done during the half stroke = the area of the piston × length of stroke × difference of mean steam pressure and atmospheric pressure.

$$= S \times l \times \left(P \frac{1 + \log. \, m}{m} - p \right)$$

$$= S\,l\,P \frac{1 + \log. \, m}{m} - S\,l\,p$$

3. The amount of steam used

$$= \text{density of steam} \times \text{space occupied by it}$$

$$= \Delta \times S \times \frac{l}{m} = \frac{S\,l}{m} \, \alpha \, P.$$

—*Translator.*

mate, for it neglects friction and the cooling due to the entry of the air."

Having, then, adopted cylinders with a diameter of 0·90 metre (2 ft. 11·43 in.), a stroke of 1·2 metre (3 ft. 11·24 in.), and the mean moment required being 3416 kilogrammetres (24,698 foot-pounds), we have—

> With an absolute pressure of 4½ atmospheres, a cut-off of ⅙th, a final pressure of 0·75 atmosphere.
> With an absolute pressure of 4 atmospheres, a cut-off of ⅕th, a final pressure of 0·8 atmosphere.
> With an absolute pressure of 3½ atmospheres, a cut-off of ¼th, a final pressure of 0·88 atmosphere.
> With an absolute pressure of 3 atmospheres, a cut-off of ⅓rd, a final pressure of 1 atmosphere.

NECESSARY CONDITIONS IN THE EXPANSION GEAR OF A WINDING ENGINE.

1. The expansion must be capable of variation. In the case of engines which are counterbalanced, the grade of the expansion must be proportional to the mean moment of the resistance. In proportion to the increase in the depth of the pit, the mean radius of the drum will increase, and the moving force must, therefore, be increased by a proportional reduction in the grade of expansion.

In the case of engines which are not counterbalanced, the grade of expansion must be varied proportionally to the resistance of the load. This variation in the grade of expansion replaces the throttling of the steam at the regulator.

A variable expansion will then become necessary in the case of engines which cannot be counterbalanced (and this is the case when the depth of the pit is more than 700 or 800 metres) in order to obtain the most economical wear of the ropes.

2. This variable expansion must be designed without complicating the engine to any great extent.

Winding engines are trusted to enginemen, neither very careful nor much instructed, and are exposed to many accidents, and must, therefore, above all things, be simple.

3. The action of the expansion gear must be altogether suspended while " changing " the cages, and at will at any point in the pit. The variation in the grade of expansion

must act automatically and without attention on the part of the engineman. In other words,' the engineman must be able to work his engine with a variable expansion with as great ease as an ordinary high-pressure engine ; and without being obliged to work any other levers than the reversing handle and the regulator.

We will now examine the different methods which have been invented for solving this problem, beginning with slide valves.

FIXED EXPANSION.—MM. Scohy and Crespin, in October, 1868, took out a patent for applying a system of expansion to winding engines. It has been described in the " Bulletin " of the old pupils of the Hainault School of Mines for April, 1870 ; and in the course of lectures on practical mining by M. Amedée Burat, in 1871. We shall only glance at the principle of this system. MM. Scohy and Crespin add to the regular form of cylinder the expansion gear of M. Meyer, as if it were an engine which worked continually in one direction ; and when the cages are being changed they do away with the effect of the expansion by admitting the steam through a supplementary passage altogether independent of the expansion gear, and which will, therefore, do away with the effect of the shutting of the expansion valve.

It is urged as an objection to this plan that it entails supplementary valves and steam ports ; that it multiplies the rubbing surfaces, which ought to be always perfectly tight, and are also never exposed to view. Apart from this objection, this system is good.

It was employed at the Monceau-Fontaine Colliery, and produced a saving of 60 per cent., according to the experiments tried by M. Scohy. This fixed expansion gear was applied to an engine where the flat iron ropes could not be counterbalanced, and they were obliged to alter the initial pressure of the steam, which was 2·90 atmospheres at the 5th turn, to 1·48 atmosphere at the 35th turn, by means of the regulator.

They were enabled, nevertheless, to alter the throttling of the steam through the regulator into the far more sensible plan of varying the expansion, by making the two slides of Meyer's expansion gear move suitably towards each other, by means of mechanism which will easily be imagined. As soon as the conditions are such that the resistance varies throughout the whole run the expansion must be made to vary automatically in proportion to the resistance.

If the complication in the machinery which results from this is feared, we must endeavour to make the moment of resistance uniform by using aloes ropes or scroll drums. The expansion gear of MM. Scohy and Crespin, which is fixed relatively to the run, is then only applicable to engines where the moment of resistance has been made constant, so as to produce its full advantage.

VARIABLE EXPANSION GEAR BY M. GUINOTTE.

The importance of altering the grade of expansion in the case of winding engines, induced M. Guinotte to produce a special design for this. Already, in 1866, M. Guinotte had taken out, in conjunction with M. Chenard, a patent for applying to winding engines a system of expansion, based on the same principle as that of MM. Scohy and Crespin. When their system was applied to the winding engine of the St. Arthur Pit, at Mariemont, M. Guinotte was induced to obtain a more complete solution of the question of expansion. M. Guinotte was enabled to produce the expansion without employing supplementary slide valves and ports, by using simply two slide valves, one behind the other; and these, moving automatically with reference to one another, produce all the grades of expansion which are required, and do away with expansion altogether during the "changing" of the cages.*

M. Guinotte published, in 1871, a remarkable paper on variable expansion. We must, however, only quote here the part which describes his system as applied to winding engines.

"My system of expansion consists in employing two slide valves superposed; the expansion valve which is fixed being

---

* It must be observed that M. Meyer's expansion gear alone, without the addition of supplementary valves and ports can never produce that instantaneous alteration from expansion to regular working which is indispensable while "changing" the cages and at any point of the run. For this alteration can only be obtained by taking a number of turns of the screw which unites the two halves of the expansion valve.

worked by a slide block, whose movement in a slide varies the admission of steam into the cylinder in all degrees from nothing to the full admission.

"In the case of winding engines, the displacement of this slide-block must necessarily be automatic, so that when the cages are being 'changed' the engine becomes of itself inexpansive, and that as soon as one cage is lifted from the bottom it becomes again expansive to the required extent. So that it is not necessary for the engineman to think at all about it, and that he may even forget that the engine is working expansively. When the run has begun, the slide-block must be in an expansive position, it must remain in that position until the cage is nearly at the surface. When the engine has only two turns more to take, the slide-block must be moved into a non-expansive position for these turns. While 'changing' it must remain in the same position, and be moved back again to the expansive position as soon as the engine has taken one or two turns in the opposite direction. To this end, it is evidently the best plan to work it from the main shaft; but whatever portion of the engine is adopted, the solutions are so easy and so numerous that each person may invent one to please himself, without difficulty; and it seems to me a waste of time to discuss them further. The plan that I have adopted myself is sufficient to demonstrate that it is quite practicable; and that is the only merit which it has to distinguish it from the others.

"But that which particularly concerns the question of varying the expansion in winding engines is another important consideration which must not be overlooked.

"It is not always possible in practice to wind the ropes on such radii that the moment of resistance should be constant throughout the run. In winding from great depths, the smallest radius will necessarily be so small, that one run would take up too many turns of the engine, and, therefore, too much time. In such cases as these it is generally the fact that a great difference between the moment of the load is tolerated or adopted so as to give a suitable radius on which to wind the ropes. This difference in the moment of the load necessitates a very defective use of the steam, for if it works well for the moment of the load of one turn, the conditions will be very bad for other turns.

"In England, they usually produce a comparative uniformity by employing a counterbalance wound upon a special roll, so as to diminish the moment throughout the first half of the run and increase it during the second half.   This plan, which is much employed in England for iron wire ropes, has produced but bad results in Belgium ; and those who have tried it there have quickly given it up.   Instead of reproducing this regularity in the moment of resistance it will perhaps be found that a better solution will be to modify, each revolution, the power exerted ; or, in other words, to make the grade of expansion vary each revolution, so that the steam should be always used in the cylinder in a proper manner.

"It is easy for anyone who has been used to handle figures to calculate in a very short time (1) the moment of the resistance for each revolution of the run (2) the amount of admission that the steam should have to the cylinder each revolution, in order that it should always work under the best possible conditions.

"From this it is easy to deduce the position which the slide-block must occupy in the slide to obtain this result.

"Let us represent (Plate XI. fig. 1) by the length A B one complete run of the engine, and divide it into as many equal parts as there are revolutions in the run.   Then draw at each of these points, which are, in fact, so many abscissæ, lines at right-angles to A B.   The lower end of the slide, which is supposed to correspond to no expansion, being represented by an ordinate equal to zero.   Then plot upon each of these perpendiculars the lengths to which the slide-block must be lifted in the slide in order to give the proper grade of expansion for each revolution.   Join each of these points by a line, which will in general be a curved line, and observe that at the end of the run we must, in all cases, come back to no expansion.   The same operation performed from the other end of A B will give us the line (partly dotted in fig. 1) which corresponds to the run of the engine in the opposite direction.

"We can now produce metallic curves corresponding to these two lines ; and by means of a shaft and screw make them reciprocate along a straight line.   The slide-block will then be fastened to a rod $t\,t$, which can only move in a vertical direction through its guides $g\,g$.

"It is kept down by its own weight, or by supplementary

weights. To this rod *t* are fastened two small levers ; the first one *l*, which is in front of the rod, can move only from left to right ; the other one *l'*, which is placed behind the rod, can move from right to left. The rod itself passes between the two metallic curves which drive it ; and these are joined together by their ends alone. The lever *l* is driven by the ·curve shown with a *full* line, the lever *l'* is driven by that which is partly dotted.

" When the metallic curves are driven in a straight line from right to left, the slide block will occupy the position corresponding to the curve shown with a full line, and when for the succeeding run it is moved back again from left to right, the slide-block will take the position corresponding to the curve shown partly with a dotted line.

" Hence, it is evident that this is a complete solution of the question of counterbalancing winding engines ; and, further, it is only to be recommended when it is not possible so to arrange the ropes, that they will produce a constant moment of resistance, or nearly so.

" It is, moreover, to be observed that my design does not necessitate any internal gear beyond that in an engine which has a fixed expansion ; and that, far from adding to the complication of Meyer's expansion gear, the internal construction is infinitely more simple.

" It is, perhaps, true that the expansion gear of MM. Scohy and Crespin, and M. Audemar, which we shall describe further on, may present, externally, greater simplicity of design, but this is in appearance alone. It matters little if we have a few levers more or less on the outside of the cylinder ; it is of very great importance not to multiply valve casings and valves, and, above all, such things as slide-valves and slide-faces, puppet-valves and valve-seats, whose condition should be almost perfect to obtain good results. However great may be the care bestowed by the builders, this perfect condition, which may be the fact, perhaps, at first, will soon become an impossibility."

Plate XI. fig. 2, represents M. Guinotte's last design for variable expansion, which was applied to the engine at "la fosse de Réunion" at the Mariemont Collieries. The gear is completely applied there ; that is to say, the moment of the load cannot possibly be made uniform, and, therefore, the power of the engine is varied so as always to maintain the same relation to the resistance of the load. Thus the period of admission of steam varies automatically, so that it amounts to one-half the stroke at the beginning of the run, and is reduced to one-tenth at the end of the run, and that there is no expansion at all when the cages are "changed."

The movement of the expansion slide-block is, in this case, obtained without any eccentric. It is obtained for one of the points x, by the piston rod and a fixed system of link work m, m_1, m_2, x; for the second point q, by the piston rod and system of link work m_1, m_3, d, q_1, which partake of the motion of the ordinary slide valve, by means of the principle K of decomposing the motion of an eccentric. The slide-block works the expansion slide, and is moved by the levers $a\,b$, $b\,c$, $c\,d$, $d\,c$, $e\,f$, and finally by the handle, $f\,g\,h$. The movement is rendered symmetrical and similar to the handle $f'\,g'\,h'$ by means of the levers $f'\,m'$, $m'\,n'$, $n'\,n$, $n\,m$, $m\,f$. These two handles are so curved that they are depressed by the two moving blocks E E′ of the Gouteaux bell signal; and make the block K move down in the curved link, and so increase the grade of expansion to the proper amount. These curved handles are so proportioned in length to the run of the cages, that when the moving blocks E and E′ get beyond their extremities, they allow the counterbalance weight l, which is placed at the end of the lever $l\,f$, to bring back the whole system of levers with which we have dealt, into the position represented in the figure in full lines, which position corresponds to the condition of *no expansion*, which is indispensable while "changing" the cages.

The engineman can, at any point in the run, make the engine work without expansion by placing his foot upon the pedal P. This design then produces in an ingenious and practical way an expansion for winding engines which varies automatically.

M. Guinotte's system of expansion has been applied successfully to five winding engines in the Mariemont Collieries. After inspecting these engines we decided to adopt it for the 250 horse-power winding engine at our "Résolu" Pit, where the depth of the pit is 665 metres (727·266 yards), and will soon be as great as 735 metres (804 yards). M. Guinotte guaranteed us an economy of 30 per cent. of coal if we applied his system of expansion to this engine, which had the ordinary slides of a high-pressure engine. Its application cost us 5000 fr. (£200) for the expansion gear complete, including erection, and 2000 fr. (£80) for the patent-right and the inventor's designs. For M. Guinotte gave detailed drawings of his system of expansion, as applied to the particular case so that each owner may be able to put up the work to tender by different firms. He superintends, moreover, the execution and erection of the gear, so that anyone who decides on adopting his gear may be certain of having it properly fitted up. We have adopted M. Guinotte's system because we consider that, of all systems which have been brought out up to this time, it is the one which produces the best results *with slide-valves* in winding engines.

This brings us to consider the last system of expansion for winding engines, that designed by M. Charles Beer.

M. CHARLES BEER'S SYSTEM OF EXPANSION.

The following is the concise description given by M. Beer of his own system.

" My design consists simply in applying the Meyer expansion gear to the winding engine ; and working the expansion by means of an *isochronous governor*.

" This governor can be set so as to give all possible degrees of expansion from zero to full admission, for a variation of about five turns between the quickest and slowest speeds of the engine.

" It is, however, to be observed that in the case of winding engines, very large variations in the grades of expansion are useless ; they will never be employed, and I think that the following may be fixed as about correct.

" 1. For high-pressure non-condensing winding engines, the expansion should vary between three-tenths and six-tenths of the stroke.

" 2. For condensing engines, the grades of expansion may vary between one-tenth and six-tenths of the stroke.

" It will be very easy to ' change ' with an admission of six-tenths ; and, between the limits which I have just given, it will be a matter of no difficulty whatever to move the two halves of the expansion valves backwards and forwards.

"Each time that the cages are 'changed' the engineman should perceptibly decrease the speed of the engine, and this will be quite sufficient to alter the expansion.

"After 'changing', when the run is begun the engineman must open fully the regulator, without thinking of the speed of the engine. The governor must produce its effect. It will keep the speed down to limits mathematically calculated beforehand. It will settle itself the grade of expansion according to the resistance to be overcome. Lastly, it will act gradually upon the steam brake, whenever the speed of the engine becomes dangerously high.

"The application of a governor to winding engines supplies them with a helper of rare intelligence, and subject to no accidents ; one which is actuated by laws which are mathematically exact, in a more certain fashion than the most skilful engineman could work.

"In the present state of mechanical science I scarcely think that it is possible to apply expansion gear to winding engines, both economically and practically, *without employing a governor*. This is the chief point to which I wish to draw attention ; and if I make mention of Meyer's system, it is because I think it a very simple one ; but many other expansion gears actuated by a governor may be employed to advantage.

"Amongst others, M. Guinotte's design, if the slide-block were actuated by the governor, would give excellent results."

In order properly to appreciate the system of M. Beer, it is necessary to consider the way in which ordinary winding engines work. We see that when "changing" the engineman admits the steam without any expansion, so as to actuate more quickly and more easily the heavy masses which have to be put in motion. These masses, in the case of an engine of 300 horse-power, are as great as 30,000 kilogs. (66,150 lbs.) without including the weight of the ropes. The full pressure of the steam is also required when one of the two pistons is on the dead point ; and it can only act on one cylinder to turn the engine round. In order to obtain the full pressure of the steam while "changing," and at any point of the run, where an engine which is not counterbalanced may be working with no steam at all, it will be necessary to give the screw, which unites the two halves of the slide of the Meyer expansion gear, a great number of turns ; and this seems to be a difficult thing to do very quickly. It is on this account, as we have said, that M. Scohy wished to add to the Meyer slide two supplementary ports and slides, so as to get, instantly, the full pressure of steam while " changing," and at any point in the run.

Let us see what happens after " changing " the cages in the case of engines which are not counterbalanced ; for, in the case of engines which are counterbalanced, where the moment of the resistance is constant, a fixed grade of expansion is sufficient.

In the case of ordinary engines the engineman opens the regulator completely to start them. On account of the inertia of the moving masses, and, consequently, of the great power required at the start, the first·turns of the engine are made slowly ; afterwards, the speed gradually increases, and from " meetings " the engineman works his engine at a high speed, in order to get the requisite mean velocity.

This great speed is brought about more easily on this account, that towards the end of the run the weight of the descending rope is so great that it will tend to draw the load up to the pulleys ; and then the engineman must shut his regulator and let the engine continue to turn, on account of the *vis viva* of the moving masses ; and during the last few turns it will even be necessary to apply back-pressure to counteract this *vis viva*.

In order to obtain a given speed of winding, we see then that it will be necessary to allow the angular velocity of the engine to increase, by employing a variable grade of expansion instead of throttling the steam through the regulator. A few figures will show more clearly what will be the conditions of the angular velocity, and the work that the steam has to do, during different turns of the run. These figures were arrived at, experimentally, by M. Scohy, in the case of an engine which was not counterbalanced, and where, therefore, the action of a fixed grade of expansion had to be supplemented by throttling the steam through the regulator, in order to properly reduce the initial pressure on the piston :—

End of Turn.	Angular Velocity.		Initial Pressure.	
	Metres.	Feet.	Atmos.	Lbs.
5th	3·38	11·093	2·90	42·630
10	4·28	14·046	2·69	39·543
15	5·06	16·607	2·48	36·456
20	5·68	18·642	2·28	33·516
25	6·28	20·610	2·05	30·135
30	6·28	20·610	1·68	24·696
35	6·28	20·610	1·48	21·756

We see, then, that the grade of expansion must increase with the angular velocity of the engine ; and that a system of variable expansion, which would tend to produce one uniform velocity, would only avoid the evil of throttling steam in the regulator, by falling into the still greater error of reducing the velocity of the engine too much.

The governor must then be calculated so as to obtain a certain number of turns of the engine, and so as only to begin to alter the speed when it exceeds this number of turns. It would be necessary, under these conditions, that the isochronous governor should make the grade of expansion alter proportionally to the resistance of the load, which is so variable in the case of engines which are not counterbalanced. It would be interesting to see this invention put to the test of absolute experiment. M. Charles Beer is at this moment applying his system of expansion to a winding engine which is nearly counterbalanced, at the Sart-Berleur Colliery, near Liége.

EXPANSION BY PUPPET VALVES.

As steam distributors, these have the following advantages over slide valves :—(1.) That of shifting the engine with less power than that required by the others, and, consequently, changing more rapidly and with greater certainty ; and this should keep pace with the increased speed of winding. In large engines, the slide valves have become so difficult to work that they have to be shifted by a special steam engine. This is the case in the pits of St. Arthur de Mariemont. (2.) That of opening more quickly the steam ports

M

to their greatest section ; this will allow us to adopt piston speeds as high as 1·80 metre, without fear of having any appreciable wire-drawing of the steam from passing through a contracted orifice. (3.) That it allows the expansion gear to produce its full effect in cutting off the steam instantaneously without appreciable wire-drawing.

It is urged as an objection to puppet valves, that they experience continual jars, which tend to destroy them, and sometimes produce an accident. For a depth of 800 metres (875 yards), the valves will experience forty-eight strokes per minute. It is urged, with a certain amount of reason, that even the working of these valves will tend to wear them out, whilst the long-continued wear of slide valves only tends to polish the rubbing surfaces (but perhaps also to scratch them through the action of the sand which is brought over by the steam). We answer to this that it is only necessary to guide the valves so as to make them always fall vertically on their seats, to make their zone of contact, which is limited, to begin with, to three or four millimetres (0·118 to 0·157 in.) hammer out and extend itself more and more. In all these cases, when the valves are worn out, they can always be replaced, as can their seats, in a few minutes.

The advantages of puppet valves are so thoroughly recognised in England and Westphalia that one sees nothing else but this form of valve. Everyone is struck with the precision, the speed, and the ease with which the engine can be turned round. Moreover, all engineers who have visited English and German mines have come back convinced of the superiority of admitting steam by means of puppet valves. Also, they have not been backward in adopting them, both in France and England. At the Paris Exhibition, Messrs. Quillacq and Co. of Anzin, showed a vertical engine worked by puppet valves. We remember, also, to have admired a horizontal winding engine exhibited by Messrs. Schneider and Co., of Creuzot, where the admission of the steam was made by means of double-beat valves whose throw was determined by two eccentrics, and the curved links of Stephenson's link motion.

All these valves were designed for an admission of high pressure steam, as in England. It remained to arrange them so as to work expansively. This improvement Mr. Audemar, engineer of the Blanzy Mines, has been the first to put in practice.

VARIABLE EXPANSION BY MEANS OF PUPPET-VALVES.

Mr. Audemar adds to the ordinary high-pressure valves, a special expansion valve which the engineman opens more or less, according to the required grade of expansion. The movement of this valve is produced by a cam with bosses, by means of a lever and a friction-roller. This cam has two motions, a rotation transmitted off the main shaft ; and a longitudinal motion along its own shaft produced by the reversing lever. The two bosses on this cam, which correspond, the one to going forwards, and the other to going backwards, are so placed that in the central position the valve is altogether shut and the engine does not go round. At both extreme positions the valve is constantly open and the engine works with no cut-off. Between these two positions, the valve is open through a greater or less portion of the stroke of the piston.

If the cam were directly coupled to the Stephenson's link, it would only be possible to obtain a high grade of expansion by approaching the link to its middle point, and, consequently, by greatly diminishing the throw of the valve, and, therefore, the opening of the steam ports. Therefore, Mr. Audemar connects the cam and the link by two toothed segments of which one, that on the cam, describes a very small angle, whilst the other describes nearly a complete circle. The result is that the steam ports maintain a very large opening even when the cam has been moved into a position to produce a high grade of expansion.

Mr. Audemar's system has been applied with success to all the heavy winding engines belonging to the Blanzy Company.

It has also been used with equal success at the Espérance collieries, at Séraing, by M. Borgnet, resident viewer, M. Godin, manager, and M. Demanet, engineer. These gentlemen have been amply recompensed by the experiment that they were the first in Belgium to make, of applying equilibrium valves to the admission of steam, and M. Audemar's valve to produce a variable expansion.

VARIABLE EXPANSION BY PUPPET VALVES,

INVENTED BY MESSRS. BRIALMONT AND KRAFT.

M. Brialmont, having to build an engine with a variable expansion for the Ougrée Colliery, considered that the special expansion valve of M. Audemar made a sort of double cut-off with the high-pressure valves, and that this would be all the more troublesome because when the Audemar valve had cut off the steam it stopped for an appreciable time between this valve and the high pressure valve. He, therefore, introduced an expansion valve worked by a cam, with bosses corresponding to each high-pressure valve.

This necessitated the use of a cam for each valve, or four cams for each cylinder. M. Kraft, when he came to design our winding engine, devised a means of working at one time the two steam valves of each cylinder and the two exhaust valves by means of two cams on one shaft, o, turned round by the engine.

We have shown on plate IX. figs. 3 and 4, this system of expansion, which is highly creditable to Messrs. Brialmont and Kraft.

As is seen in fig. 3, each cam, n and m, has two bosses, one for turning forwards and one backwards. The form of these bosses regulates the opening of the steam and exhaust valves. Fig. 4 shows that the shaft o turns the cam n, and makes it lift, by means of the friction-roller g, and the levers l and p, the steam valve s'. The form of the boss, n, then lets the roller g make the valve s' drop upon its seat, and cut off the steam instantaneously, so as to allow the expansion to produce its full effect without appreciable wire-drawing. The shaft o, as it continues to rotate, makes the cam n lift the roller g', and works, by means of the lever f, the other steam valve so as to produce the back stroke of the piston. As for the exhaust valve, it is clear that at the same time that the cam n opens the steam valve, the cam m works the roller *r* and, through the levers *f* and *g*, lifts the exhaust valve *s*.

If the engineman moves the cams longitudinally along the shaft and places them in *mid-gear*, instead of in the position represented in fig. 3, he then stops all steam from going *in* or *out*. If he places the cams in their extreme position or *full gear*, it admits the steam without any expansion.

By means of one lever only, that for shifting the cams, the engineman can then make the engine turn a head or turn backward; and, in either direction, he can make it work either with the full pressure of steam or any grade of expansion he pleases.

According to this system, then, the two eccentrics and Stephenson's link, which are used for each of the two cylinders of the engines as usually arranged, are altogether done away with.

The engine can now be turned round wonderfully easily. At the Ougrée Colliery, where the regulator was kept entirely open, and where the engine worked with a vacuum, the engineman could make the crank turn backwards and forwards by a simple movement of the cams. It is to be observed that when, instead of moving four cams for each cylinder the engineman need only move two cams, the engine can by this be turned round much more easily. Nothing can be conceived more simple, more easy, and more mechanical than this arrangement of expansion gear and valves. The most remarkable point about it is that it induces the engineman not to touch his regulator at all, but to produce all possible movements of the engine, by moving one lever alone, including the change in the direction of motion, and also quickening or slowing the speed of the engine. The engineman will not be tempted to touch a second lever, that which works the regulator, unless an accident should happen to some of the gear for the admission of steam. It will be an inevitable consequence, that for throttling of the steam through the regulator will be substituted *working expansively*, the most rational and economical way of working that has been designed.

Is it a valid objection to this system that by it the expansion cannot be varied automatically? Certainly not, since we can with certainty obtain a constant moment of resistance by using rope-rolls and aloes ropes, or, again, by spiral drums so constructed as to counterbalance round steel ropes.*

It is, therefore, necessary to employ one fixed grade of expansion throughout nearly all the run, and place the lever which works the cams into the notch corresponding to this grade. It is only towards the end of the run that the engine-man need increase the grade of expansion, then stop all admission of steam, and let the cage be lifted to the bank by the *vis viva* of the moving parts.

According to the Seraing system of variable expansion, it is possible also to produce, automatically, a variation of the expansion proportional to the variation in the resistance, by replacing the cams by a piece of mechanism analogous to that designed by M. Guinotte for his slide valves or piston valves.

It is to be observed that it is of great importance so to counterbalance the ropes as not to necessitate a great change in expansion, for, otherwise, a great part of the power of the engine is wasted.

In that case the friction inherent to an engine working with a vacuum, becomes of great importance, particularly with high speed ; and that the useful effect of the engine is reduced in proportion. M. Burat quotes on this subject the engine of Sainte Marie de Montceau, which works expansively and developes 277 indicated horse-power, measured by a Watt's indicator, to produce only a useful effect in the pit of 92 horse-power, that is to say, only 33 per cent. of the former. However, when the engine of the Saint Pierre Pit was made to work at full power they obtained a useful effect varying from 64 to 59 per cent. with a grade of expansion from one-half to one-third. In order that the resistance, although it be considerable, should be in proportion to the size of the cylinders, the grade of expansion must not be varied to any great extent. Moreover, if the expansion is to be varied automatically, it will always be necessary to couple up the machinery, and this will complicate the engine.

We see, then, that the best winding engines, that is to say,

---

* It is to be remarked that spiral drums enable us to get a great speed of winding with but a small piston speed, and above all with but a small number of turns of the engine, and this is of the greatest possible advantage in actual work. Of course, it necessitates larger cylinders.

those which have great simplicity in the machinery, and also great economy of ropes and coal, have (1) scroll drums for round steel ropes ; (2) an engine with expansion gear, of fixed or moveable grade at will, and with the Seraing valves.*

POSITION OF WINDING ENGINES.

The small distance which there is at our colliery between the winding engine and the pit, as well as the confined space which we have for the engines, have made us adopt engines with vertical cylinders. It is a recognised fact that vertical engines possess the following advantages :—

· 1. They avoid the unequal wear of cylinders and piston rods, whereas, in spite of the long stuffing-boxes and guides which are employed at the back end of horizontal cylinders, it is impossible to prevent the slides from being much worn after several years, and the weight of the pistons from wearing unequally the cylinders and making them oval; and this necessitates heavy expenses when they have to be replaced.

2. They enable the drum-shaft to be lifted up so high as to diminish the inclination of the ropes considerably.

3. They enable the engineman to be nearer to the pit-top, and this is of the greatest importance to prevent the possibility of the cages being over-wound. When the speed of winding is as high as from 10 to 15 metres per second (10·936 to 16·404 yards) the slightest inattention or the slightest confusion in the signals might cause a cage to go over the pulleys at a speed which nothing could withstand. It is well known that good safety hooks will hold up the cage, but they allow the rope to be hurt. When, however, the engineman is so near to the pit-top that he is, as it were, close upon it, he can work his engines with a speed and safety which are much

---

* If we consider the figures given by M. Burat of the annual cost in coal (from 40,000 to 100,000 fr.) necessitated by winding 100,000 tons of coal, we shall see how frightfully costly it is, in fuel, to adopt cylindrical drums or rope-rolls for steel ropes in winding from a great depth; on account of the excessive variation in power required, resulting from the large diameters of the drums which are necessary in order to preserve the ropes. The increase in the cost of fuel will then largely exceed the annual saving of from 2700 to 5000 fr., which will result from employing round or flat steel ropes instead of aloes ropes.

greater than is possible when he is 30 metres off (32·809 yards), as is the case with horizontal engines.

Unfortunately, vertical engines, by necessitating the drum shaft to be so far removed from the foundation on the ground level, require two heavy parallel walls to prevent its being drawn towards the pit top ; and these strengthened by enormous counter-forts.   These heavy foundations hide from the engineman's view the valves and valve-gear, when he stands at the level of the drum shaft.   This objection is got over in the case of horizontal engines, which give to the whole framing the greatest possible steadiness by placing the drum shaft on the ground, and enabling the engineman to see the whole of his gear.

In order to make these grave objections to vertical engines of as little importance as possible, M. Kraft has carried his drum shaft upon a group of cast-iron columns, without any masonry walls whatever; so as to enable the engineman to see all his gear at the same time ; and has tied the head gear to this cluster of columns by an enormous wrought-iron box girder.

This tying of the head gear to the winding engine, which we are employing at the Grand-Mambourg Colliery, near Liége, is shown on Plate IX.   It represents an immense framing where all the strains from moving forces and resistances are placed in equilibrium and neutralise each other.

For the weights hanging in the pit transmit their pressure to the pulley axle, o, by means of the ropes $n\,p$ and $n\,m$; and tend to draw this axle towards the pit.   We may resolve the resultant of these forces into two forces, as follows :—A vertical force v tending to lift up the drum-shaft, which is counteracted by the weight of the drums, and the ties and bolts which tie the framing down to the foundations.   The other force h, in a horizontal direction from the entablature of the engine, and tending to pull it over towards the pit, that is to say, towards the end a of the girder $a\,b$.   The winding engine, on the other hand, exerts a force by means of the ropes $n\,p$ and $n\,m$, which tends to draw the head gear towards the engine.   The girder $a\,b$ is placed there to resist this thrust and makes it re-act upon the cluster of columns which carry the drums.

We may split up this force again into two, one V acting

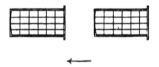

vertically and counteracted by the resistance of the columns and foundations to a crushing strain; the other, H, acting horizontally along the entablature of the engine, which tends to counteract the thrust h towards the pit, which we have mentioned before.

We see, then, that the effect of the two forces acting at the points $a\,b$ may be resolved into (1) their horizontal components meeting in the entablature $m\,a$, and neutralising each other; (2) two vertical components easily resisted by the foundations. It is, therefore, evident that it will be no longer necessary to carry the drum shaft on blocks of masonry, but that a cluster of cast-iron columns which expose to view the whole engine will answer as well. An iron head gear consisting of two vertical lattice girder legs and two struts, is unusually simple. The fact that it will not catch fire is an advantage which can scarcely be over-rated, considering the fires which have recently destroyed the wooden head gears at Mont-Saint Aldegonde, and at Sebastopol Pit, at Trieu-Kaisen, and have caused such disastrous consequences. .

The plan of tying the head gear and the winding engine together is altogether both bold and rational; it is the design of M. Kraft.

The numerous improvements which have been made in different details of the engines, drum, steam brake, steam valves and gear worked by the engineman, have been due, together, to M. Kraft and M. Deney, the engineer at the Cockerill Works, who had charge of the drawings for these engines.

We have entered at so great length into the recent progress that has been made in winding engines, that we cannot spend time over these questions of detail. Moreover, the drawing of the engine will supply these.

BOILERS GROUPED ROUND THE FURNACE.

We cannot, however, leave off without mentioning our system of boilers.

On the score of economy of coal, questions relative to boilers are of the highest importance in colliery work, if one

N .

considers that the ranges of boilers supply more than 500 horse-power, and that it is possible to realise a very great saving by having boilers which will allow of a good combustion, and a greater or less absorption of the *radiant* and *conducted* heat resulting from it.

The author adopted, with this view, the system of coupled boilers designed by M. Paul Havrez, and detailed in the *Revue des Mines* in 1863.

According to this plan, two heaters round a furnace are coupled up to one ordinary boiler, and this enables them to absorb the radiant heat as perfectly as Cornish boilers, and at the same time gets rid of the objections to these, viz., imperfect combustion, large diameter, thick plates, weight, costliness, complication, etc., and gives them, as well, the advantages of boilers consisting of heaters and reheaters, without the corresponding disadvantages, viz., good combustion and simple and cheap construction, without many costly fire-bricks, and imperfect absorption of the radiant heat. Our line of boilers, which has now worked successfully for four years, gives out more than 300 horse-power.

The collieries of Poirier and Mont-sur-Marchienne have also just adopted it. It has produced the following results : —During an experiment which lasted seven and three-quarter hours, and under a pressure of from two and a-half to three and a-half atmospheres, we have produced 10·56 kilogs. of steam with 1 kilog. of coal, and 17 kilogs. of steam per square metre of heating surface per hour (31·337 lbs. per square yard). A long series of experiments has produced six results, consistent with each other, and with the best possible conditions for combustion and utilisation of the heat developed. The Charleroi section will be invited to examine this system of boilers, and it will also be made the subject of a special paper.*

* In the first series of experiments, which lasted four hours, and was made before a deputation consisting of Messrs. Depoitier, mining engineer, and Maroquin, chief engineer of the Couillet and Châtelineau works, it was proved that 1 kilog. of coal evaporated 10·49 kilogs. of water at a pressure of from three and a-quarter to two and a-quarter atmospheres (10·49 lbs. of water with 1 lb. of coal), and that 1 square metre of heating surface evaporated 18 kilogs. of water per hour (33·18 lbs. of water per square yard). It remains to work out the results for a pressure of three and a-half atmospheres.

The capital results given by this arrangement of boilers come from the following facts :—

1. It admits of a perfect combustion.

2. It enables the plates to absorb all the radiant heat, and to carry off almost immediately all the flame from this heat, so that the combustible gases can become cooled over a comparatively small boiler surface. The small quantity of heat which is wasted round the furnace and in the chimney, affords an explanation of its extraordinary economy.

CONCLUSIONS.

In designing new sets of winding machinery we must, therefore, employ the following :—

1. M. Lambert's system of pit guides.

2. Cages and trams built of mild steel.

3. Round ropes made of mild steel, tapering as far as possible, and, in order to preserve them and counterbalance them when at work, spiral drums in which the pitch of the helix and the size of the groove increase proportionally to the distance of each point from the vertical plane through the pulley.

4. High-pressure winding engines, with a high grade of expansion, either fixed or adjustable at will, and with the Seraing valves.

5. The coupled boilers of M. Paul Havrez, which are economical both in fuel and in the first cost of the apparatus.

APPENDIX A.

SIZES, WEIGHTS, AND STRENGTHS OF ROPES OF VARIOUS MAKERS.

MESSRS. R. S. NEWALL & Co.

ROUND WIRE ROPES.

Iron.		Steel.		Newall's improved steel.		Equivalent strength.		
Circum. In.	Lbs. per fathom.	Circum. In.	Lbs. per fathom.	Circum. In.	Lbs. per fathom.	Working load for pits. Cwt.	Working load for inclines. Cwt.	Breaking strain. Tons.
1	1	—	—	—	—	4	6	2
$1\frac{1}{2}$	$1\frac{1}{2}$	1	1	—	—	6	9	3
$1\frac{5}{8}$	2	—	—	—	—	8	12	4
$1\frac{3}{4}$	$2\frac{1}{2}$	$1\frac{1}{2}$	$1\frac{1}{2}$	1	1	10	15	5
$1\frac{7}{8}$	3	$1\frac{5}{8}$	2	$1\frac{1}{2}$	$1\frac{1}{2}$	12	18	6
2	$3\frac{1}{2}$	—	—	—	—	14	21	7
$2\frac{1}{8}$	4	$1\frac{3}{4}$	$2\frac{1}{2}$	$1\frac{5}{8}$	2	16	24	8
$2\frac{1}{4}$	$4\frac{1}{2}$	—	—	—	—	18	27	9
$2\frac{3}{8}$	5	$1\frac{7}{8}$	3	—	—	20	30	10
$2\frac{1}{2}$	$5\frac{1}{2}$	2	$3\frac{1}{2}$	$1\frac{3}{4}$	$2\frac{1}{2}$	22	33	11
$2\frac{5}{8}$	6	—	—	—	—	24	36	12
$2\frac{3}{4}$	$6\frac{1}{2}$	$2\frac{1}{8}$	4	$1\frac{7}{8}$	3	26	39	13
$2\frac{7}{8}$	7	—	—	—	—	28	42	14
3	$7\frac{1}{2}$	$2\frac{1}{4}$	$4\frac{1}{2}$	—	—	30	45	15
$3\frac{1}{8}$	8	$2\frac{3}{8}$	5	2	$3\frac{1}{2}$	32	48	16
$3\frac{1}{4}$	$8\frac{1}{2}$	—	—	—	—	34	51	17
$3\frac{3}{8}$	9	$2\frac{1}{2}$	$5\frac{1}{2}$	$2\frac{1}{8}$	4	36	54	18
$3\frac{1}{2}$	10	$2\frac{5}{8}$	6	—	—	40	60	20
$3\frac{5}{8}$	11	$2\frac{3}{4}$	$6\frac{1}{2}$	$2\frac{1}{4}$	$4\frac{1}{2}$	44	66	22
$3\frac{3}{4}$	12	$2\frac{7}{8}$	7	$2\frac{3}{8}$	5	48	72	24
$3\frac{7}{8}$	13	3	$7\frac{1}{2}$	$2\frac{1}{2}$	$5\frac{1}{2}$	52	78	26
4	14	$3\frac{1}{8}$	8	$2\frac{5}{8}$	6	56	84	28
$4\frac{1}{4}$	15	$3\frac{3}{8}$	9	—	—	60	90	30
$4\frac{3}{8}$	16	$3\frac{1}{2}$	10	$2\frac{3}{4}$	$6\frac{1}{2}$	64	96	32
$4\frac{1}{2}$	18	$3\frac{5}{8}$	11	3	$7\frac{1}{2}$	72	108	36
$4\frac{3}{4}$	20	$3\frac{3}{4}$	12	$3\frac{1}{8}$	8	80	120	40
5	22	$3\frac{7}{8}$	13	$3\frac{3}{8}$	9	88	132	44
$5\frac{1}{2}$	26	$4\frac{1}{4}$	15	$3\frac{5}{8}$	11	104	156	52
6	33	$4\frac{3}{4}$	20	4	14	132	198	66

Messrs. R. S. Newall & Co.

(Continued).

FLAT WIRE ROPES.

Iron.		Steel.		Newall's improved steel.		Equivalent strength.	
Size in inches.	Lbs. per fathom.	Size in inches.	Lbs. per fathom.	Size in inches.	Lbs. per fathom.	Working load. Cwt.	Breaking strain. Tons.
$2\frac{1}{4} \times \frac{1}{2}$	10	—	—	—	—	32	16
$2\frac{1}{2} \times \frac{1}{2}$	12	—	—	—	—	40	20
$2\frac{3}{4} \times \frac{5}{8}$	14	—	—	—	—	48	24
$3\frac{3}{4} \times \frac{5}{8}$	16	$2\frac{1}{4} \times \frac{1}{2}$	10	—	—	56	28
$3\frac{1}{4} \times \frac{5}{8}$	18	$2\frac{1}{2} \times \frac{1}{2}$	12	—	—	64	32
$3\frac{1}{2} \times \frac{5}{8}$	20	$2\frac{3}{4} \times \frac{5}{8}$	14	$2\frac{1}{4} \times \frac{1}{2}$	10	72	36
$3\frac{3}{4} \times \frac{11}{16}$	22	$2\frac{3}{4} \times \frac{5}{8}$	14	$2\frac{1}{4} \times \frac{1}{2}$	10	80	40
$4\frac{3}{4} \times \frac{11}{16}$	25	$3 \times \frac{5}{8}$	16	$2\frac{1}{2} \times \frac{1}{2}$	12	88	44
$4\frac{1}{4} \times \frac{3}{4}$	28	$3 \times \frac{5}{8}$	—	—	—	96	48
$4\frac{1}{2} \times \frac{3}{4}$	31	$3\frac{1}{4} \times \frac{5}{8}$	18	$2\frac{3}{4} \times \frac{5}{8}$	14	104	52
$4\frac{5}{8} \times \frac{3}{4}$	34	$3\frac{1}{4} \times \frac{5}{8}$	—	—	—	112	56
$4\frac{7}{8} \times \frac{7}{8}$	36	$3\frac{1}{4} \times \frac{5}{8}$	20	$3\frac{3}{4} \times \frac{5}{8}$	16	120	60
—	—	$4\frac{1}{2} \times \frac{3}{4}$	31	$3\frac{1}{2} \times \frac{5}{8}$	20	160	90

	Iron rope. Cwt.	Steel. Cwt.	Newall's improved steel. Cwt.
Working load for inclines per lb. per fathom	... 6	... 10	... 14
" pits "	... 4	... 7	... 10
Breaking strain "	... 40	... 70	... 100

Hartlepool Patent Ropery Company.

ROUND WIRE ROPES.

Hemp.		Charcoal iron wire.		Steel.		Equivalent strength.	
Circumference.	Lbs. per fathom.	Circumference.	Lbs. per fathom.	Circumference.	Lbs. per fathom.	Working load. Cwt.	Breaking strain. Tons.
2¾	2	1	1	—	—	6	2
—	—	1½	1½	1	1	9	3
3¾	4	1⅝	2	—	—	12	4
—	—	1¾	2½	1½	1½	15	5
4½	5	1⅞	3	—	—	18	6
—	—	2	3½	1⅝	2	21	7
5½	7½	2⅛	4	1¾	2½	24	8
—	—	2¼	4½	—	—	27	9
6	9	2⅜	5	1⅞	3	30	10
—	—	2½	5½	—	—	33	11
6½	10	2⅝	6	2	3½	36	12
—	—	2¾	6¾	2⅛	4	39	13
7	12	2⅞	7	2¼	4½	42	14
—	—	3	7½	—	—	45	15
7½	14	3⅛	8	2⅜	5	48	16
—	—	3¼	8½	—	—	51	17
8	16	3⅜	9	2½	5½	54	18
—	—	3½	10	2⅝	6	60	20
8½	18	3⅝	11	2¾	6¾	66	22
—	—	3¾	12	—	—	72	24
9½	22	3⅞	13	3¼	8	78	26
10	26	4	14	—	—	84	28
—	—	4¼	15	3⅜	9	90	30
11	30	4⅜	16	—	—	96	32
—	—	4½	18	3½	10	108	36
12	34	4⅝	20	3¾	12	120	40

FLAT WIRE ROPES.

Hemp.		Charcoal iron wire.		Patent steel.		Equivalent strength.	
Size in inches.	Lbs. per fathom.	Size in inches.	Lbs. per fathom.	Size in inches.	Lbs. per fathom.	Working load. Cwt.	Breaking strain. Tons.
4 × 1	16½	2⅛ × ½	10	—	—	35	16
4½ × 1⅛	20	2½ ×	12	—	—	40	18
5 × 1¼	24	2¾ × ⅝	14	1⅞ × ½	8	50	23
5½ × 1⅜	26	3 ×	16	2⅛ ×	10	60	27
6 × 1½	28	3¼ ×	18	2½ ×	12	70	32
7 × 1⅞	36	3½ × ¾	20	2¾ × ⅝	14	80	36
8¼ × 2⅛	40	4 ×	25	3 ×	16	100	40
8½ × 2¼	45	4⅞ × ⅞	35	3½ × ¾	20	110	45

Haggie Brothers.

ROUND ROPE.

Tarred hemp.		White manilla.		Charcoal iron wire.		Steel wire.		Equivalent strength.	
Circum. In.	Lbs. per fathom.	Circum. In.	Lbs. per fathom.	Circum. In.	Lbs. per fathom.	Circum. In.	Lbs. per fathom.	Working load. Cwt.	Breaking strain. Tons.
2	1	2	½	1	1	—	—	6	2
2½	2	2½	1	1½	1½	1	1	9	3
3½	4	3	1½	1¾	2½	—	—	15	5
4½	6	3¾	2½	2	3½	1⅝	2	21	7
5½	8	4½	3	2¼	4½	1⅞	3	27	9
6	9	4¾	3¼	2½	5½	2	3½	33	11
6½	11	5	3¾	2¾	6½	2¼	4½	39	13
7	12	6	5¼	3	7½	2⅜	5	45	15
7½	14	6¼	5½	3¼	8½	—	—	51	17
8	16	6½	6	3½	10	2⅝	6	60	20
8½	18	6¾	6½	3¾	12	3	7½	72	24
9	20	7	7	4	14	3¼	8	84	28
10	26	7½	8½	4¼	16	—	—	90	30
11	30	8½	11	4½	18	3½	10	108	36
12	36	9½	14	5	21	4	14	115	40
14	48	9¾	15	5½	26	—	—	128	50
16	64	10	16	6	32	—	—	130	65
—	—	11	18	—	—	—	—	—	—

FLAT ROPE.

Hemp.				Charcoal wire.		Steel wire.		Equivalent strength.	
Breadth four strands.	Lbs. per fathom.	Breadth six strands.	Lbs. per fathom.	Breadth.	Lbs. per fathom.	Breadth.	Lbs. per fathom.	Working load. Cwt.	Breaking strain. Tons.
3½ in.	10	4 in.	10	2¼ in.	11	—	—	44	20
4	13½	4¾	13½	2½	13	—	—	52	23
4½	17	5½	17½	2⅝	14	—	—	56	24
5	22	6	20	2¾	15	—	—	60	27
5¼	23½	6¼	22¼	3	16	2 in.	10	64	29
5½	25½	6½	24½	—	—	2¼	11	67	30
5¾	28	6¾	26½	—	—	2½	13	70	31
6	30	7	28	3¼	18	—	—	72	32
6½	36	7½	32	3½	20	2¾	15	80	36
7	41½	8¼	39	4	25	2⅞	15½	100	45
7½	48	9	47	4¼	28	3	16	112	50
8	55	9½	52	4½	32	3¼	18	128	56
8½	65	10	58	4¾	34	—	—	136	60
9	72	11	70	5¼	40	3½	20	146	70

Glaholm and Robson.

ROUND ROPES.

Hemp.		Combination.	Iron.		Steel.		Equiv. strength.	
Circum. In.	Lbs. per fathom.	Lbs. per fathom.	Circum. In.	Lbs. per fathom.	Circum. In.	Lbs. per fathom.	Working Load. Cwt.	Brkng. strain. Tons.
2¾	2	1¾	1	1	—	—	5	1½
3	2½	2	1¼	1¼	—	—	6	2¾
3¼	3	2¼	1½	1½	1	1	8	3
3¾	4	3¼	1⅝	2	1¼	1¼	10	4
4	4½	3¾	1¾	2½	1½	1½	12	4¾
4½	5	4	1⅞	3	—	—	15	6
5	6	5	2	3½	1⅝	2	17	7
5¼	7	5¾	2⅛	4	—	—	20	7½
5½	8	6¾	2¼	4½	1¾	2½	22	8
6	8½	7	2⅜	5	1⅞	3	24	9½
6¼	9	7¾	2½	5½	—	—	26	10
6½	10	8	2⅝	6	2	3½	28	11
6¾	11	9	2¾	6½	2⅛	4	30	12
7	12	9¾	2⅞	7	2¼	4½	33	14
7¼	12½	10	3	7½	2⅜	5	36	15
7½	13	10¾	3⅛	8	—	—	38	16
7¾	14	11½	3¼	8½	—	—	40	17
8	15½	12	3⅜	9½	2½	5½	43	18
8¼	16	12	3½	10	2⅝	6	47	18½
8½	17	12¾	3⅝	11	2¾	6½	51	19
9	19	14¼	3¾	12	3	7½	60	20½
9½	22	16	3⅞	13	3⅛	7¾	63	24
10	26	21	4	14	3¼	8	67	28
10¼	27	22	4⅛	14½	—	—	70	30
10¾	28	22¾	4¼	15	3⅜	9	76	32
11	30	23½	4⅜	16	—	—	80	34
11½	32	25	4½	18	3½	10	88	36
12	34	27	4¾	20	3¾	12	95	40
12½	35½	28	4⅞	21½	3⅞	13	100	42
13	37	29½	5	22½	4	14	110	44

FLAT ROPES FOR PITS.

Russian hemp.			Combnd. hemp.		Iron.		Steel.		Equiv. stgth.	
Breadth. In.	Lbs. per fm.		Lbs. per fm.		Breadth. In.	Lbs. per fathom.	Breadth. In.	Lbs. per fathom.	Working load. Cwt.	Breaking strain. Tons.
	4 std.	6 std.	4 std.	6 std.						
4	14½	—	12½	—	2½	13	—	—	35	14
4¼	17	—	15	—	2⅝	14	—	—	40	16
4½	18½	—	16	—	2¾	15	—	—	50	18
5	22	—	19½	—	—	—	—	—	—	—
5½	26	—	23½	—	2⅞	15	2	10	54	20
6	31	—	28	—	3	16	2¼	11	60	26
5	—	17	—	14½	3½	20	2½	13	70	33
5¾	—	19	—	16½	—	—	—	—	42	16
6	—	23	—	20	—	—	—	—	50	20
6½	—	26	—	22½	—	—	—	—	53	21
—	—	—	—	—	—	—	—	—	65	26
—	—	—	—	—	3¾	22	2¾	15	80	37
—	—	—	—	—	4	25	3	16	90	42
—	—	—	—	—	4¼	28	3¼	18	110	50
—	—	—	—	—	4½	32	3½	20	130	58

Craven & Speeding Brothers.

ROUND ROPES.

Hemp.		Charcoal iron wire.			Steel.		Equivalent strength.	
Circumference. In.	Lbs. per fathom.	Dia. Inches.	Cir. Inches.	Lbs. per fathom.	Circumference. In.	Lbs. per fathom.	Working load. Cwt.	Breaking strain. Tons.
2	1	5/16	1	1	—	—	2	1
2½	2	½	1½	2	1	1	4	2
3½	4	9/16	1¾	3	—	—	8	4
4½	6	⅝	2	4	1⅝	2	12	6
5½	8	11/16	2¼	5	1⅞	3	15	8
6	9	13/16	2½	6	2	3½	18	10
6½	11	⅞	2¾	7	2¼	4½	22	12
7	12	1	3	8	2⅜	5	25	14
7½	14	1 1/16	3¼	9	—	—	30	16
8	16	1⅛	3½	10	2⅝	6	35	18
8½	18	1 3/16	3¾	12	3	7½	38	20
9	20	1¼	4	14	3¼	8	40	22
10	26	1⅜	4¼	16	—	—	45	28
11	30	1 7/16	4½	18	3½	10	50	34
12	36	1⅝	5	21	4	14	55	40
14	48	1¾	5½	26	—	—	65	54
16	64	1⅞	6	32	—	—	75	70

FLAT ROPES.

Hemp.				Charcoal wire.		Steel wire.		Equivalent strength.	
Breadth four strands. In.	Lbs. per fathom.	Breadth six strands. In.	Lbs. per fathom.	Breadth. In.	Lbs. per fathom.	Breadth. In.	Lbs. per fathom.	Working load. Cwt.	Breaking strain. Tons.
3½	10	4	10	—	—	—	—	14	—
4	13½	4¾	13½	2½	13	—	—	20	16
4½	17	5½	17½	2⅝	14	—	—	22	17
5	22	6	20	2¾	15	—	—	25	18
5¼	23½	6¼	22¼	3	16	2	10	30	21
5½	25½	6½	24½	—	—	2¼	11	38	24
5¾	28	6¾	26½	—	—	2½	13	40	26
6	30	7	28	3¼	18	—	—	45	28
6½	36	7½	32	3½	20	2¾	15	50	31
7	41½	8¼	39	4	25	2⅞	15½	55	34
7½	48	9	47	4¼	28	3	16	60	37
8	55	9½	52	4½	32	3¼	18	70	40
8½	63	10	58	4¾	—	—	—	85	46
9	72	11	70	5¼	—	3½	20	120	60

Messrs. Wilkins and Weatherly.

STEEL AND CHARCOAL IRON WIRE ROPES.

Patent steel wire.		Iron wire.		Equivalents in hemp.			
Circumference. In.	Lbs. per fathom.	Circumference. In.	Lbs. per fathom.	Circumference. In.	Lbs. per fathom.	Working load. Cwt.	Breaking strain. Tons.
4¼	16	6	27	13	37	160	48
4	14	5½	23	12	33	135	41
3⅝	11¾	5	20	11½	31	120	36
3½	11	4⅝	18	11	30	105	33
3⅜	9½	4¼	16	10½	29	96	29
3⅛	8½	4	14	10	28	84	25
3	8	3¾	13	9½	25	78	23
2⅞	7	3⅝	11¾	9	22	70	21
2¾	6	3½	11	8½	20	66	19
2⅝	5¾	3⅜	9½	8	16	57	17
2½	5	3⅛	8½	7½	14	50	15
—	—	3	8	7	12	45	14
2¼	4	2⅞	7	6½	10	42	13
2⅛	3¾	2¾	6	6	9	36	11
2	3¼	2⅝	5¾	5½	8	34	10
1⅞	3	2½	5	5	7	28	9
1¾	2½	2¼	4	4½	6	24	7
1⅝	2	2⅛	3¾	4	5	22	6½
—	—	2	3¼	3¾	4½	20	6
1½	1¾	1⅞	3	3½	4	18	5
1⅜	1½	1¾	2½	3	3½	15	4
1¼	1¼	1½	1¾	2¾	3	10	3
1⅛	1	1⅜	1½	2½	2½	8	2½
—	—	1¼	1¼	2¼	2	6	2

FLAT ROPES.

Patent steel wire.		Iron wire.		Equivalents in hemp.			
Size in inches.	Lbs. per fathom.	Size in inches.	Lbs. per fathom.	Size in inches.	Lbs. per fathom.	Working load. Cwt.	Breaking strain. Tons.
4⅝ × ⅞	34	6 × 1	59	13 × 3	98	210	85
4¼ × 11/16	28	5½ × 1	48	11½ × 2¾	78	175	70
3¾ × ¾	24	5 × ⅞	39	10 × 2½	63	145	58
3½ × 11/16	21	4½ × ⅞	32	8½ × 2¼	52	120	48
3¼ × ⅝	18	4¼ × 13/16	28	7½ × 2⅛	45	105	42
3 × ⅝	16	4 × ¾	26	7 × 1⅞	42	92	37
2⅞ × 9/16	14	3¾ × ¾	24	6½ × 1⅝	39	82	33
2¾ × ½	12	3½ × 11/16	21	6 × 1½	34	72	29
—	—	3¼ × ⅝	18	5¾ × 1½	29	60	24
2 × ⅝	11	3 × ⅝	16	5½ × 1⅜	26	52	21
1⅞ × ⅝	9	2⅞ × 9/16	14	5¼ × 1¼	24	45	18
—	—	2¾ × ½	12	5 × 1¼	22	40	16
—	—	2 × ⅝	11	4 × 1⅛	20	35	14

DIXON, CORBITT, AND SPENCER.

ROUND ROPE.

Hemp.		Iron.		Steel.		Ex. strg. steel.		Equiv. strength.	
Circum. In.	Lbs. per fathom.	Circum. In.	Lbs. per fathom.	Circum. In.	Lbs. per fathom.	Circum. In.	Lbs. per fathom.	Workg. load. Cwt.	Breakg. strain. Tons.
$2\frac{3}{4}$	2	1	1	—	—	—	—	6	2
—	—	$1\frac{1}{2}$	$1\frac{1}{2}$	1	1	—	—	9	3
$3\frac{3}{4}$	4	$1\frac{5}{8}$	2	—	—	—	—	12	4
—	—	$1\frac{3}{4}$	$2\frac{1}{2}$	$1\frac{1}{2}$	$1\frac{1}{2}$	1	1	15	5
$4\frac{1}{2}$	5	$1\frac{7}{8}$	3	$1\frac{5}{8}$	2	$1\frac{1}{2}$	$1\frac{1}{2}$	18	6
—	—	2	$3\frac{1}{2}$	—	—	—	—	21	7
$5\frac{1}{2}$	7	$2\frac{1}{8}$	4	$1\frac{3}{4}$	$2\frac{1}{2}$	$1\frac{5}{8}$	2	24	8
—	—	$2\frac{1}{4}$	$4\frac{1}{2}$	—	—	—	—	27	9
6	9	$2\frac{3}{8}$	5	$1\frac{7}{8}$	3	—	—	30	10
—	—	$2\frac{1}{2}$	$5\frac{1}{2}$	2	$3\frac{1}{2}$	$1\frac{3}{4}$	$2\frac{1}{2}$	33	11
$6\frac{1}{2}$	10	$2\frac{5}{8}$	6	—	—	—	—	36	12
—	—	$2\frac{3}{4}$	$6\frac{1}{2}$	$2\frac{1}{8}$	4	$1\frac{7}{8}$	3	39	13
7	12	$2\frac{7}{8}$	7	—	—	—	—	42	14
—	—	3	$7\frac{1}{2}$	$2\frac{1}{4}$	$4\frac{1}{2}$	—	—	45	15
$7\frac{1}{2}$	14	$3\frac{1}{8}$	8	$2\frac{3}{8}$	5	2	$3\frac{1}{2}$	48	16
—	—	$3\frac{1}{4}$	$8\frac{1}{2}$	—	—	—	—	51	17
8	16	$3\frac{3}{8}$	9	$2\frac{1}{2}$	$5\frac{1}{2}$	$2\frac{1}{8}$	4	54	18
—	—	$3\frac{1}{2}$	10	$2\frac{5}{8}$	6	—	—	60	20
$8\frac{1}{2}$	18	$3\frac{5}{8}$	11	$2\frac{3}{4}$	$6\frac{1}{2}$	$2\frac{1}{4}$	$4\frac{1}{2}$	66	22
—	—	$3\frac{3}{4}$	12	$2\frac{7}{8}$	7	$2\frac{3}{8}$	5	72	24
$9\frac{1}{2}$	22	$3\frac{7}{8}$	13	3	$7\frac{1}{2}$	$2\frac{1}{2}$	$5\frac{1}{2}$	78	26
10	26	4	14	$3\frac{1}{8}$	8	$2\frac{5}{8}$	6	84	28
—	—	$4\frac{1}{4}$	15	$3\frac{3}{8}$	9	—	—	90	30
11	30	$4\frac{3}{8}$	16	$3\frac{1}{2}$	10	$2\frac{3}{4}$	$6\frac{1}{2}$	96	32
—	—	$4\frac{1}{2}$	18	$3\frac{5}{8}$	11	3	$7\frac{1}{2}$	108	36
12	34	$4\frac{3}{4}$	20	$3\frac{3}{4}$	12	$3\frac{1}{8}$	8	120	40

FLAT ROPES.

Iron.		Steel.		Ex. strong steel.		Equiv. strength.	
Size in inches.	Lbs. per fathom.	Size in inches.	Lbs. per fathom.	Size in inches.	Lbs. per fathom.	Working load Cwt.	Breaking strain Tons.
$2\frac{1}{4} \times \frac{1}{2}$	10	—	—	—	—	40	18
$2\frac{1}{2} \times \frac{1}{2}$	12	—	—	—	—	48	20
$2\frac{3}{4} \times \frac{5}{8}$	14	—	—	—	—	56	24
$3 \times \frac{5}{8}$	16	$2\frac{1}{4} \times \frac{1}{2}$	10	—	—	64	28
$3\frac{1}{4} \times \frac{5}{8}$	18	$2\frac{1}{2} \times \frac{1}{2}$	12	—	—	72	32
$3\frac{1}{2} \times \frac{5}{8}$	20	$2\frac{3}{4} \times \frac{5}{8}$	14	$2\frac{1}{4} \times \frac{1}{2}$	10	80	36
$3\frac{3}{4} \times \frac{11}{16}$	22	$2\frac{3}{4} \times \frac{5}{8}$	14	$2\frac{1}{4} \times \frac{1}{2}$	10	88	40
$4 \times \frac{11}{16}$	25	$3 \times \frac{5}{8}$	16	$2\frac{1}{2} \times \frac{1}{2}$	12	100	45
$4\frac{1}{4} \times \frac{3}{4}$	28	$3 \times \frac{5}{8}$	—	—	—	112	50
$4\frac{1}{2} \times \frac{3}{4}$	31	$3\frac{1}{4} \times \frac{5}{8}$	18	$2\frac{3}{4} \times \frac{5}{8}$	14	128	56
$4\frac{5}{8} \times \frac{3}{4}$	34	$3\frac{1}{4} \times \frac{5}{8}$	—	—	—	136	60
$4\frac{7}{8} \times \frac{7}{8}$	36	$3\frac{1}{4} \times \frac{5}{8}$	20	$3 \times \frac{5}{8}$	16	150	64

George Cradock.

ROUND WIRE ROPE.

ROPES OF EQUIVALENT STRENGTH.									
Improved steel.		Ordinary steel.		Charcoal iron.		Hemp.			
Circum. In.	Lbs. per fathom.	Circum. In.	Lbs. per fathom.	Circum. In.	Lbs. per fathom.	Circum. In.	Lbs. per fathom.	Brkng. strain. Tons.	Wrkng. load. Cwt.
—	—	¾	1	—	—	3¼	2½	2	6
—	—	1¼	1¼	1½	2	3½	3	2½	8
¾	1	1⅜	1½	1¾	2½	4¼	4	4	13
1	—	1½	2	1⅞	3	4½	4¾	4¼	15
				2	3½	4¾	5½	5	17
1⅜	1½	1¾	2½	2⅛	4	5	6	5½	18½
				2¼	4½	5¼	6½	6	20
1½	2	1⅞	3	2⅜	5	5½	7¼	6½	22
		2	3½	2½	5½	5¾	8	7	24
1¾	2½			2⅝	6	6	8¾	7¾	27
		2⅛	4	2¾	6½	6¼	9½	8½	30
				2⅞	7	6¾	11	11	36
1⅞	3	2¼	4½	3	7½	7	12	12	40
				3⅛	8	7½	13	13	44
—	—	2⅜	5	3¼	8½	7⅝	14	14	47
		2½	5½	3⅜	9½	8	16	15	50
2	3½	2⅝	6	3½	10½	8½	18½	16	54
				3⅝	11¼	8¾	19	17	57
2⅛	4	2¾	6½	3¾	12	9	20	18	62
2¼	4½	2⅞	7	3⅞	13	9¼	21	19	66
2⅜	5	3	7½	4	14	9½	22	20	70
2½	5½	3⅛	8	4⅛	14¾	9¾	24	21½	75
2⅝	6	3¼	9	4¼	15½	10	26	23	80
2¾	6½	3⅜	10	4⅜	16¾	10¼	27	24½	85
3	7½	3½	11	4½	18	10½	28	26	90
3⅛	8	3⅝	11½	4⅝	19	11	30	28	95
		3¾	12	4¾	20	11½	33	30	100
3¼	9	3⅞	13	4⅞	21	12	36	33	105
		4⅛	14¾	5	25½	Wire strand hearts in these sizes.			110
3⅜	10	4¼	15½	5¼	27				115
3½	11	4⅜	16¾	5½	28				120
3⅝	11½	4½	18	5¾	30½				125
3¾	12	4⅝	19	6	34				130

FLAT WIRE ROPE.

ROPES OF EQUIVALENT STRENGTH.							
Ordinary steel.		Charcoal iron.		Hemp.			
Size in inches.	Lbs. per fathom.	Size in inches.	Lbs. per fathom.	Size in inches.	Lbs. per fathom.	Working load. Cwt.	Breaking strain. Tons.
1⅞ by ½	8	2⅛ by ½	13	4½ by 1¼	17	30	10
2 ,, 9/16	10	2¾ ,, ½	15	4¾ ,, 1 5/16	19	40	13
2⅛ ,, 9/16	12	3 ,, 9/16	16	5 ,, 1⅜	21	44	16
2½ ,, ½	13	3¼ ,, ⅝	18	6¼ ,, 1¼	23	52	20
		3½ ,, ⅝	20	7¼ ,, 1⅜	30	60	20
2¾ ,, ½	15	3¾ ,, 11/16	22	8 ,, 1½	38	70	29
3 ,, 9/16	16	4 ,, ¾	25	8½ ,, 1⅝	42	80	34
		4¼ ,, ¾	28	9¼ ,, 1¾	48	90	38
3⅛ ,, ⅝	18	4⅜ ,, 13/16	31	10 ,, 2	58	100	42

Mr. John Shaw, Jun.

Russian hemp.		Best cast steel.		Best selected charcoal wire rope.			
Circumference.	Lbs. per fathom.	Circumference.	Lbs. per fathom.	Circumference.	Lbs. per fathom.	Breaking strain. Tons.	Working load. Cwt.
3½	3	1¼	1½	1½	2	2¾	9
4	4	1⅜	1¾	1¾	2¾	4	15
4½	5	1½	2	2	3¼	6	20
5	6½	1¾	2¾	2¼	4½	7½	24
5½	7½	1⅞	3	2½	5½	9½	30
6	8½	2	3¼	2¾	6½	11½	36
6½	10	2¼	4½	3	7¾	14	45
7	12	2½	5½	3¼	8	16	52
7½	14	2⅝	6	3½	10	19	62
8	16	2¾	6½	3¾	12	22	74
8½	18	3	7¾	4	14	25	80
9	20	3¼	8½	4¼	16	28	95
9½	22	3⅜	9½	4½	18	32	105
10	24	3¾	12½	4¾	20	36	120
10½	28	4	14½	5	22	40	135
11	30	4⅛	15	5¼	25	45	150
11½	33	4⅜	17	5½	28	50	160
12	37	4½	18	5¾	31	55	170
13	46	4¾	20	6	35	60	180

Flat hemp rope.		Flat steel rope.		Flat charcoal iron wire rope.			
Size in inches.	Lbs. per fathom.	Size in inches.	Lbs. per fathom.	Size in inches.	Lbs. per fathom.	Breaking Strain. Tons.	Working load. Cwt.
3½ by 1	12	—	—	2¼ by ½	10	18	40
4 „ 1¼	15	—	—	2½ „ ½	12	20	45
4½ „ 1½	20	—	—	2¾ „ ⅝	14	23	51
5 „ 1¾	24	2¼ by ½	10	3 „ ⅝	16	27	56
5½ „ 1¾	27	2½ „ ½	12	3¼ „ ⅝	18	30	60
6¼ „ 1⅞	30	2¾ „ ⅝	14	3½ „ ⅝	20	33	68
6½ „ 2	33	3 „ ⅝	16	3¾ „ ¾	22½	36	76
7 „ 2	36	3¼ „ ⅝	18	4 „ ¾	25	39	96
7½ „ 2⅛	39	3½ „ ½	21	4¼ „ ⅞	28	42	105
8 „ 2¼	42	3¾ „ ¾	22½	4½ „ ⅞	32	45	120

W. M. Hutchings, Printer, 5, Bouverie Street, and 38, Wilderness Lane,
Dorset Street, E.C.

CPSIA information can be obtained
at www.ICGtesting.com
Printed in the USA
LVHW011354111218
600055LV00004B/554/P